DATABASES IN SYSTEMATICS

Proceedings of an International Symposium held in Southampton

THE SYSTEMATICS ASSOCIATION
SPECIAL VOLUME NO. 26

DATABASES IN SYSTEMATICS

Edited by

R. ALLKIN
and
F. A. BISBY

Biology Department, The University, Southampton, England

1984

Published for the
SYSTEMATICS ASSOCIATION
by
ACADEMIC PRESS

LONDON ORLANDO SAN DIEGO SAN FRANCISCO NEW YORK
TORONTO MONTREAL SYDNEY TOKYO SÃO PAULO

ACADEMIC PRESS INC. (LONDON) LTD.
24–28 Oval Road
London NW1 7DX

U.S. Edition published by
ACADEMIC PRESS INC.
(Harcourt Brace Jovanovich, Inc.)
Orlando, Florida 32887

British Library Cataloguing in Publication Data

Databases in systematics.—(The Systematics
Association special volume, ISSN 0309-2593; 26)
1. Information storage and retrieval systems—
Biology—Classification 2. Data base
management
I. Allkin, R. II. Bisby, F.A. III. Series
574'.028'5442 QH83

ISBN 0-12-053040-6

Printed in Great Britain at the Alden Press
Oxford, London and Northampton

Contributors

Adey, M. E., *Biology Department, Building 44, The University, Southampton SO9 5NH, UK*

Allkin, R., *Biology Department, Building 44, The University, Southampton SO9 5NH, UK*

Babaç, M. T., *Biyoloji Bölümü, Fırat Universitesi, Elazıg, Turkey*

Barron, D. W., *Computer Studies Department, The University, Southampton SO9 5NH, UK*

Bisby, F. A., *Biology Department, Building 44, The University, Southampton SO9 5NH, UK*

Bocquet, G., *Conservatoire et Jardin botaniques, Case postale 60, CH-1292, Chambesy, Geneva, Switzerland*

Brill, R. C., *University of Michigan Computing Center, 1075 Beal Avenue, Ann Arbor, MI 48109, USA*

Charlwood, B. V., *Department of Plant Sciences, King's College London, 68 Half Moon Lane, London SE24 9JF, UK*

Crovello, T. J., *Department of Biology, The University of Notre Dame, Notre Dame, Indiana 46556, USA*

Dadd, M. N., *BIOSIS UK Limited, 44 High Street, Boston Spa, Wetherby, West Yorkshire LS23 6EA, UK*

Dallwitz, M. J., *CSIRO Division of Entomology, P.O. Box 1700, Canberra City 2601, Australia*

Derrick, L. N., *Plant Science Laboratories, Botany Department, The University, Whiteknights, Reading RG6 2AS, UK*

Estabrook, G. F., *University of Michigan Herbarium, 2000 North University Building, Ann Arbor, MI 48109, USA*

Farquhar, G., *Computer Centre, Strathclyde University, 204 George Street, Glasgow G1 1XW, UK*

Feoli, E., *Istituto ed Orto Botanico, Cas. Università, I 34100, Trieste, Italy*

Flesness, N. R., *ISIS, 12101 Johnny Cake Ridge Road, Apple Valley, Minnesota 55124, USA*

Freeston, M. W., *Computer Studies Department, The University, Southampton SO9 5NH, UK*

Gama, L., *Instituto Nacional de Investigaciones sobre Recursos Bióticos, Apdo. Postal No. 63, Xalapa, Ver., Mexico*

Garnatz, P. G., *Computer Software Associates Inc., P.O. Box 8089, Roseville, Minnesota 55113, USA*

Gibbs Russell, G. E., *Botanical Research Institute, Private Bag X101, Pretoria, Republic of South Africa*

Gómez-Pompa, A., *Instituto Nacional de Investigaciones sobre Recursos Bióticos, Apdo. Postal No. 63, Xalapa, Ver., Mexico*

Gonsalves, P., *Botanical Research Institute, Private Bag X101, Pretoria, Republic of South Africa*

Green, F. N., *Agricultural Scientific Services, Department of Agriculture and Fisheries for Scotland, East Craigs, Edinburgh EH12 8NJ, UK*

Grenham, M. J., *Department of Plant Sciences, King's College London, 68 Half Moon Lane, London SE24 9JF, UK*

Hauser, L. A., *Department of Botany, The University of Illinois, Urbana, Illinois 61801, USA*

Heywood, V. H., *Plant Science Laboratories, Botany Department, The University, Whiteknights, Reading RG6 2AS, UK*

Keller, C. A., *Department of Education, Andrews University, Berrien Springs, Michigan, USA*

Kelly, M. C., *BIOSIS Inc., 2100 Arch Street, Philadelphia, PA 19103-1399, USA*

Lucas, G. Ll., *Royal Botanic Gardens, Kew, Richmond, Surrey TW9 3AB, UK*

Macfarlane, T. D., *Western Australian Herbarium, George Street, South Perth, W.A. 6151, Australia*

Mackinder, D. C., *IUCN Conservation Monitoring Centre, c/o The Herbarium, Royal Botanic Gardens, Kew, Richmond, Surrey TW9 3AB, UK*

Margot, P., *Federal Institute of Technology, Lausanne, Switzerland*

Mascherpa, J. M., *Conservatoire et Jardin botaniques, Case postale 60, CH-1292 Chambesy, Geneva, Switzerland*

Mitchell, K. A., *Plant Science Laboratories, Botany Department, The University, Whiteknights, Reading RG6 2AS, UK*

Moore, D. M., *Plant Science Laboratories, Botany Department, The University, Whiteknights, Reading RG6 2AS, UK*

Moreno, N. P., *Instituto Nacional de Investigaciones sobre Recursos Bióticos, Apdo. Postal No. 63, Xalapa, Ver., Mexico*

Morris, G. S., *Clinical Sciences Laboratories, Guy's Hospital, London Bridge, London SE1 RT, UK*

Nimis, P. L., *Istituto ed Orto Botanico, Cas. Università I, 34100, Trieste, Italy*

Pankhurst, R. J., *Botany Department, British Museum (Natural History), Cromwell Road, London SW7 5BD, UK*

Pignatti, S., *Istituto di Botanica, Citta Universitaria, 00100 Roma, Italy*

Scheepen, J. van, *Plant Science Laboratories, Botany Department, The University, Whiteknights, Reading RG6 2AS, UK*

Seal, U. S., *U.S.V.A. Hospital, Mpls., Minnesota 55455, USA*

Sieg, J., *Universität Osnabrück/Abteilung Vechta, Fachbereich, Naturwissenschaften, Mathematik, Driverstraße 22, D-2848 Vechta, Federal Republic of Germany*

Sosa, V., *Instituto Nacional de Investigaciones sobre Recursos Bióticos, Apdo. Postal No. 63, Xalapa, Ver., Mexico*

Watling, R., *Royal Botanic Garden, Inverleith Row, Edinburgh EH4 3HU, UK*

White, R. J., *Biology Department, Building 44, The University, Southampton SO9 5NH, UK*

Winfield, P. J., *Agricultural Scientific Services, Department of Agriculture and Fisheries for Scotland, East Craigs, Edinburgh EH12 8NJ, UK*

Preface

December 1982 proved to be an excellent date for the Systematics Association to hold an international symposium on "Databases in Systematics". The feeling that we were experiencing a period of rapid technological development, particularly in the effectiveness of small computers and the availability of database software, was matched by a fresh impetus to bring this technology to the many information handling tasks in systematics. This book contains the 26 papers given at the symposium.

Computer databases are not new to systematics: an important, sometimes sobering, part of the symposium was to review the successes and difficulties of the very earliest projects. The symposium heard how nomenclatural, biogeographic and curatorial data had been successfully handled in computer databases for some time. It was also valuable to hear of both managerial and human aspects of these projects as well as to look at developments planned for the future.

The organizing committee chose to emphasize the recent interest in databases handling descriptive biological data. Whilst descriptive data has frequently been held in computers for individual tasks such as cluster analysis, on-line identification or printing descriptions, there is now a rapid development of exploratory multi-purpose descriptive databases. The latter part of the symposium was devoted to papers on this theme: on software development, the logical arrangement of data, the particular difficulties of descriptive data, and to a range of experimental descriptive database projects involving floristic or monographic approaches, morphological, chemical or genetic data, and taxa ranging from animal families to plant cultivars.

The design of systematic databases raises several questions which are of fundamental importance to systematists regardless of their interest in technological applications. One of these is the nature of descriptive information. What are the essential elements of observations, comparisons, taxonomic characters, descriptions and diagnoses? Another is the form of the information services provided by taxonomists either for the public or for use within the profession. Who are the customers, what types of information are needed, how can the information most readily be made available?

The result was a stimulating and creative exchange of ideas: participants came from sixteen countries and the principal research teams in this field all made contributions. The sessions ran smoothly thanks to the chairmen – Professor R. G. Davies (President of the Association), Professor D. M. Moore (Past President), Professor T. J. Crovello, Mr R. J. Pankhurst and Mr G. Ll. Lucas, to whom we are most grateful.

We are very much indebted to Mr R. J. Pankhurst who joined us on the organizing committee and to Dr R. J. White for organizing the demonstrations. We are most grateful to Mrs Margaret Newton for the work she did as symposium secretary and to Mrs J. Rihan, Mr N. Maxted and Mr P. Hillier for their invaluable help behind the scenes.

We acknowledge with thanks the contributions made by the Royal Society, the British Council and the Systematics Association towards the expenses of speakers coming from overseas.

R. Allkin
December 1983 *F. A. Bisby*

Contents

Contributors v

Preface ix

1 Electronic Data Processing in Taxonomy and Systematics
V. H. HEYWOOD 1

2 Information Services in Taxonomy
F. A. BISBY 17

3 Current Database Design – The User's View
D. W. BARRON 35

4 The Implementation of Databases on Small Computers
M. W. FREESTON 43

5 Management of Almost Flat Files in Systematic Biology using TAXIR
R. C. BRILL and G. F. ESTABROOK 53

6 A Concept for a Machine-readable Taxonomic Reference File
M. N. DADD and M. C. KELLY 69

7 The European Taxonomic, Floristic and Biosystematic Documentation System – An Introduction
V. H. HEYWOOD, D. M. MOORE, L. N. DERRICK, K. A. MITCHELL and J. van SCHEEPEN 79

8 The Database of the IUCN Conservation Monitoring Centre
D. C. MACKINDER 91

9 ISIS – An International Specimen Information System
N. R. FLESNESS, P. G. GARNATZ and U. S. SEAL 103

10 The Network of Databanks for the Italian Flora and Vegetation
P. L. NIMIS, E. FEOLI and S. PIGNATTI 113

11 Fact Documentation and Literature Database for the Crustacean Order Tanaidacea
J. SIEG 125

12 PRECIS – A Curatorial and Biogeographic System
G. E. GIBBS RUSSELL and P. GONSALVES 137

13 A Review of Herbarium Catalogues
R. J. PANKHURST 155

14 Flora of Veracruz: Progress and Prospects
A. GOMEZ-POMPA, N. P. MORENO, L. GAMA, V. SOSA and R. ALLKIN 165

15 The Vicieae Database: An Experimental Taxonomic Monograph
M. E. ADEY, R. ALLKIN, F. A. BISBY, T. D. MACFARLANE and R. J. WHITE 175

16 The Use of a Descriptive Database as an Aid to Assessing the Distinctness of Pea Cultivars (*Pisum sativum* L.)
P. J. WINFIELD and F. N. GREEN 189

17 A Chemical Database for the Leguminosae
B. V. CHARLWOOD, G. S. MORRIS and M. J. GRENHAM 201

18 A Chemotaxonomic Database
M. T. BABAÇ and F. A. BISBY 209

19 BRASS BAND (The Brassicaceae Data Bank at Notre Dame): An Example of Database Concepts in Systematics
T. J. CROVELLO, L. A. HAUSER and C. A. KELLER 219

20 An Outline for a Database within a Major Herbarium
J. M. MASCHERPA and G. BOCQUET 235

21 Identification of Toxic Mushrooms and Toadstools (Agarics) – An On-Line Identification Program
P. MARGOT, G. FARQUHAR and R. WATLING 249

22 Handling Taxonomic Descriptions by Computer
R. ALLKIN 263

23 Automatic Typesetting of Computer-generated Keys and Descriptions
M. J. DALLWITZ 279

24 Implementing Small Database Systems with Specialized Features
R. J. WHITE 291

25 On the Description of Inflorescences
R. J. PANKHURST 309

26 Databases in Systematics: A Summing Up
G. Ll. LUCAS 321

Index of Key Words 325

List of Systematics Association Publications 330

1 Electronic Data Processing in Taxonomy and Systematics

V. H. HEYWOOD

*Department of Botany, University of Reading,
Reading RG6 2AS, UK*

Abstract: Taxonomists have paid scant regard to the effectiveness of the information-processing side of their subject until recent years. Many of the procedures of taxonomy are archaic and wasteful of labour so that the effectiveness of the world's small workforce is more limited than it need be. The introduction of electronic data processing (EDP) techniques to taxonomic practice has been slow in comparison with the position in other disciplines. This stems partly from the ambiguous nature of taxonomy as a science (or non-science). There are, however, signs of a change in attitude as lessons are being learned from the experience gained by using EDP in other scientific and non-scientifc areas. Widespread use of EDP is essential if taxonomy is to meet some of the time-related goals now being set for it.

INTRODUCTION

Most branches of science, and indeed many other aspects of human endeavour, have been radically transformed in the period following World War II. This is not surprising when one considers the dramatic developments in instrumentation and technology which occured during that period, leading in biology, for example, to rapid, cheap and efficient methods of chemical analysis and so making possible the now flourishing fields of phytochemistry and biochemical systematics; likewise electron microscopy (transmission, scanning, scanning-transmission and high voltage) has come of age in a way few could

Systematics Association Special Volume No. 26, "Databases in Systematics", edited by R. Allkin and F. A. Bisby, 1984, pp. 1–15. Academic Press, London and Orlando.

DATABASES IN SYSTEMATICS
ISBN 0 12 053040 6

have predicted. But it is the electronic revolution, involving computers, microprocessors, and accompanying sophisticated software and increased storage capacity that has probably had the greatest overall effect on science, as a rapid tour of the laboratories of any scientific institution will show.

The impact of electronic data processing in the various fields of taxonomy and systematics will, I believe, come to be looked upon, in retrospect, as one of the most curious episodes in the history of biology. It is well over ten years since the Flora North America programme was put forward as an information system (Shetler, 1971, 1973; Krauss, 1973) and already by 1973, wide-eyed optimism had been replaced by red-eyed realism. Shetler, one of the main architects of FNA found himself constrained to present a paper at the first International Congress for Systematic and Evolutionary Biology (ICSEB) entitled "Demythologizing Biological Data Banking" in which he asserts, with a backdrop of several years of "hands-on" experience, that biologists have been quite unrealistic in their predictions of what the computer can do for them in the realm of computerized data banking. Of course, many biologists would disagree with that assessment. Shetler's paper makes depressing reading and would cause any systematist interested in possible applications of EDP to steer well clear were it not for the fact that the situation has changed dramatically in the past eight to ten years. Shetler's paper was given in a symposium entitled "Computer Revolution in Systematics" but the convenor Mello (1974) noted that whether the computer was to be regarded as a revolutionary device was uncertain at that time. I believe we can be more positive today, largely because of the experience we have obtained from the application of EDP in other fields that can be applied to systematics.

The application of EDP to taxonomy and systematics can be considered under three main headings:

(1) data management or information processing, i.e. housekeeping;
(2) classification, i.e. clustering techniques, and identification and key-making (together comprising numerical taxonomy/taximetrics);
(3) cladistic analysis and evolutionary reconstruction.

This chapter will concentrate on information processing, for this is the realm of systematics where the computer has been slowest to invade. As Shetler (1974) comments, this is surprising when one considers the colossal information retrieval problem that the taxonomist has always faced, although he suggests that the lag may be a reflection of "the constitutional aversion of the scientist to slowing up the pace of primary discovery for the sake of organizing and synthesizing what is already discovered". Although I would accept that this is part of the explanation, it is also, I believe, an attitude that reflects a philosophical dichotomy that has characterized taxonomy since its establishment as a science in the mid-eighteenth century. So to place EDP in context, this historical dimension will be explained below.

HISTORICAL CONSIDERATIONS – SCIENCE VS. SCHOLASTICISM

It is generally considered that scientific taxonomy was established by Linnaeus. I have argued on previous occasions that Linnaeus in fact marked the end of an era, not the beginning of a new one (Heywood, 1980, 1984). His major achievement was one that had more to do with information processing than science: the codification of the traditional folk and herbalist taxonomies as recorded in the chaotic pre-Linnaean literature. He also introduced a technical language for describing plants, removed verbs, and reduced descriptions to a stylized, regimented set of statements in which properties typical of the living plant were rejected. In a sense, Linnaeus did not so much describe plants as subjugate the description to taxonomic ends (Jacobs, 1980). Additionally, Linnaeus is credited with having invented the loose-leaf as opposed to the bound-volume herbarium, allowing the sheets (on which only a single species was mounted) to be sorted, rearranged and expanded like a genuine archive. The descriptive terminology, stylized descriptions, establishment of the herbarium technique and consistent application of the binomial system provided taxonomy with the means to deal with the world's flora. What is not often appreciated is that this also marked what Jacobs (1980) calls alienation of botany from everyday life, as well as the divorce of plant taxonomy from the living world of nature.

In practice, Linnaeus replaced a system of folk taxonomies which were readily comprehended by the societies in which they developed by a codified classification of so esoteric a nature that it required the creation of the profession of taxonomists to apply it to the world's flora that was rapidly being explored. This resulted in an increase of species, genera and higher groupings on a scale that Linnaeus could never have envisaged. In short, Linnaeus created a methodology but did not have to live with it. He was, by nature, more of an information scientist than a taxonomist. If he were alive today, he would probably be seduced by the "curatorial syndrome" of Shetler (1974) for he was first and foremost an organizing genius, despite his undoubted powers of description. As Lindroth (1978) noted:

> The urge for completeness and order, the talent for describing vast amounts of empirical facts and putting them into neat systems . . . has always been a characteristic of Swedish intellectual life.

Linnaeus performed these feats of organizing genius at a time when men of science and learning were beginning to reject the Aristotelean basis on which the Linnaean edifice was built, in favour of a rational, scientific approach which was concerned not just with empirical description but with seeking explanations and interpretations, proceeding beyond classification to understanding nature. I refer to the prophets of the age of enlightenment and the encyclopaedists, such as Buffon, Lamarck, Diderot, and the first empirical scientist taxonomist, Adanson.

Here we see the beginnings of the dichotomy in taxonomy between its role of information processing for biology and as a branch of experimental or, at least, explanatory science. We can also see why, due to the sheer amount of new material that had to be handled as the results of the voyages of exploration, taxonomic decisions became more important than the information that led to them, for the decision, through the binomial, was what was quoted, with an authority (the word is used deliberately).

DECISION MAKING IN TAXONOMY

The disinterest of taxonomists in the efficiency of the methodology employed to construct a general systematic classification of the world's organisms, has meant that much of the information obtained about these organisms during the various classificatory processes is then lost by being subordinated to the taxonomic decisions and not recorded (Raven, 1977). This is what I have called elsewhere the shoe-box syndrome of taxonomy (Heywood, 1974). Decision making, in fact, dominates taxonomy: Floras and other taxonomic publications, for example, are collections of decisions or opinions. Decision-making is a very complex process. In most floristic works, no provision is made for indicating what evidence has been used for making decisions apart from the limited morphological data (and occasional chromosome number) given. This is highlighted by Taylor (1971) who comments that once we have made our decisions on what taxa to recognize

> we tried to organize all our information to support this decision and end by making the assumption we have made our most important contribution by making the taxonomic decision itself. Clearly, much of the information about the organism is lost in such a process because in the final communication, we have transmitted only an abstraction premised by a taxonomic decision and have lost the goal of constructing a bank of information about the organism.

In short, we organize our information so as to justify the decisions we have arrived at!

Worse, there may be little connection between the data involved in the decision-making process and the communication via description (Heywood, 1973). One of the reasons for the lack of interest shown by taxonomists in the information-processing side of their subject is the ambiguous nature of taxonomy as a science – whether it is scientific in Popper's terms or, indeed, whether taxonomy is of such a nature as to invalidate the Popperian conception of science (cf. Gray, 1980).

The widespread belief that taxonomy is not scientific has been renewed recently by Hennigian cladists and various subcults, e.g. transformed cladists such as Patterson, Platnick and others, and has led taxonomists to seek respectability in various ways: through the meticulous scholarship of systematic monographs, the search for phylogenetic or evolutionary relationships in post-Darwinian times, the study of population variation, structure and micro-evolution in biosystematic and experimental taxonomy and so on.

INFORMATION PROCESSING

> If we are truly interested in cataloguing this diversity . . . it seems all but incomprehensible that we have not made use of electronic data processing to keep track of the units being classified and allow the much more efficient accumulation of information about them. There is literally no other way in which this can be done, and the fact that funds have not been appropriated to make it possible suggests to me that society at large does not assign a particularly high priority to the task of cataloguing organic diversity.
>
> P H Raven (1977)

One of the main factors to be taken into account in considering the use of EDP in taxonomy is, of course, the size of the potential computerized database. It is not, therefore, surprising to find that most initiatives in this area have been taken by smaller institutions or by individuals, usually university-based. It is self-evident that, as a research project, a small database can be constructed with relatively little commitment of resources or personnel, whereas a large-scale database requires a major policy decision and change of operational procedures of radical and long-term proportions. In general, the larger the institution, the greater the commitment to traditional, existing procedures and, consequently, the greater the inertia to be overcome.

After much prompting, an international conference was held in 1973 at Kew, England, sponsored by the Eco-Sciences Panel of NATO, to review the ways in which EDP methods could be used in the major European taxonomic collections (Brenan *et al.*, 1975). One of the outcomes of the meeting was the setting up of a Working Party to take the matter further, which was divided into three groups: one to consider descriptors, another systems and software, and the third a type-register. The report of the Descriptors Group identified three categories of essential descriptors:

(1) *curatorial*, such as Herbarium code, accession number, collector name, etc;
(2) *taxonomic and nomenclatural* such as family, generic names, specific epithets, type indicator, etc; and
(3) *geographical* such as country, geographical coordinator, locality name, altitude, etc.

Morphological descriptors (which are discussed elsewhere in this symposium) pose a different set of problems and were not considered in detail.

This group also considered examples of the ways in which the data/descriptors could be processed, such as monographic listings of the specimens classified taxonomically, floristic listings for a given area or country, compilation of botanical gazeteers, lists of endangered species, automatic map-plotting and various curatorial lists.

Subsequent progress in implementing EDP applications in major herbaria, either in Europe or elsewhere, has been disappointingly slow. Nearly ten years after the Kew meeting we can point to only a few substantial examples. This is due, more, I suspect, to a combination of the inbuilt inertia referred to earlier and the resistance to international cooperation between major institutions, than to technical difficulties. All taxonomic activity forms part of an international network of information and communication. Although individual pieces of research can be, and are, undertaken in apparent isolation, all taxonomy is dependent on a series of internationally agreed conventions regarding names, publications, taxonomic structure and even the basic units involved both in terms of categories and actual named taxa. Throughout the world, taxonomists consult *Index Kewensis*, *Engler's Syllabus*, the *International Code of Botanical Nomenclature*, *Index Herbariorum*, etc. The same styles and formats of Floras and monographs are used across the world, following one of several standard sets of abbreviations of authors, journals, etc. There are several direct methods of cooperation between institutions, such as the system of loan of materials for writing revisions, monographs or, in some cases, Floras. This is a labour-intensive and very expensive procedure, for both lending and receiving institutions, and in some cases loans are now highly restricted or even discontinued on financial grounds.

Yet despite this kind of inter-institutional cooperation, there is virtually no coordination on fundamental issues, such as the completion of the floristic inventory – in other words, agreement on priorities so that who does what is clearly established. Of course, there is some degree of cooperation through large-scale Flora projects such as the recently initiated "Flora Meso-Americana" undertaking, involving institutions in Mexico, the United States and the United Kingdom, but apart from the "International Association for Plant Taxonomy" (which tends to confine its activities to nomenclatural and bibliographical matters) there is no organization or body responsible for global coordination of resources in plant or animal taxonomy. Several approaches have been made but have proved fruitless and moves are afoot to try and establish a mechanism for cooperation at curator level.

Following a detailed study, the European Science Foundation's Taxonomy Group included the following recommendations in its final report *Taxonomy in Europe* (Heywood and Clark, 1982):

> We urge that a meeting of representatives of the main European taxonomic centres be convened (possibly under the auspices of the ESF and of IUBS) to review the current situation in plant taxonomy in the light of recent developments, and to prepare an overall strategy in association with institutions in the United States and other world centres of taxonomy.

The picture revealed by the Group's investigation is of "a more or less chaotic and incoherent, uncoordinated mixture of large and small projects with no overall aim or theme".

There is no authoritative world survey of floristic projects in progress and no one centre to which one can turn for information. The *Guide to Standard Floras of the World*, recently published by Frodin (1984) as a purely personal venture and produced by Cambridge University Press, represents a major achievement and will form an exceedingly valuable basis for the creation of an appropriate databank – a necessity if the work is not to become outdated in a few years, especially since so many of the Floras listed are in active preparation. What is needed is not just a continually updated inventory of inventories (Floras), so to speak, but some way of monitoring progress.

Herbaria and museums constitute an indispensible archive for systematic biology and it is important that much greater efforts should be made to facilitate access to the collections by means of catalogues and indexes. Indexes of classical collections, type-materials and, in appropriate cases, microfiches (as exist already for a very few collections such as the Linnaean herbarium) would reduce the pressure on loans, although by no means would solve it. Other desirable procedures include the circulation of lists of identification between Institutions holding duplicates of important recent collections, and the extension of this to more routine identifications as suggested by Cullen (1984) so as to avoid duplicated effort. Storage of the basic information in a databank, with the facility for easy updating and circulation of the information by tapes or on-line as appropriate, is an obvious approach already being adopted in a number of all-too-few instances.

The report of the St Louis Workshop on Trends, priorities and needs in systematic and evolutionary biology (Anon., 1974a) noted that:

> Systematics and evolutionary biology is now faced with staggering problems of information storage and retrieval, and of the adequate preservation of the specimens and literature that have been gathered over so many years at such a high cost.

The question of costs is an important one. For too long we have proceeded in systematic biology without regard to the background costs: it has been assumed as self-evident that governments will continue indefinitely to maintain the national collections in herbaria and museums and, indeed, provide funds for their expansion. Just how parlous the financial support is for some major collections was highlighted in the Preliminary Report of the European Science Foundation Taxonomy Working Group entitled *Taxonomy in Europe* (European Science Foundation, 1977). The matter was referred by the Group to the International Committee of Natural History Museums (ICNHM) of the International Commission on Museums and debated at the ICNHM meeting in Mexico City in 1980. ICNHM resolved to establish a Working Group to examine and coordinate existing information on collection and future curatorial needs, especially those pertinent to the conservation of the world's natural resources. I have no knowledge of subsequent action, and action it is that is needed.

What has to be stressed, however, is that escalating costs coupled with a lack of clearly defined goals, is liable to lead to a gradual collapse of the system. It is no longer obvious that, for example, the United Kingdom government will continue indefinitely to support two major world herbaria in London (Kew and British Museum) or that the Swedish government will continue likewise to maintain the geographically close major herbaria at Stockholm (over 4 000 000 specimens) and Uppsala (over 2 000 000), not to mention Lund (also with 2 000 000 specimens). There are such nonsenses as Manchester Herbarium with its worldwide collection of some 3 000 000 specimens and practically no staff or activity, and in France, the Herbarium at the University of Lyon which with nearly 4 000 000 sheets is virtually unstaffed.

Valuable information on the running costs of the US Botanical Systematic Collections is given in a two-part report, *Systematic Botany Resources in America*, to the National Science Foundation (Anon., 1974b, 1979). Relevant to this present discussion were replies in answer to the question: "Is current funding from all sources sufficient to maintain the herbarium in all of its aspects?" Sixty percent replied "No", including one federal facility; 70% believed that a teletype network among major herbaria could be a useful tool for systematic botany; and a remarkable 91% responded "Yes" to the question "Do you believe a central data bank would be a useful tool for systematic research?", although only four curators out of a total of 623 who replied (504 did not!) listed EDP support as being a critical need facing their institutions. Moreover, only 20 out of the 623 institutions that replied indicated that they were involved in EDP programmes of their own or were participating in programmes using EDP. These ranged from placing label-data from herbarium sheets on punch cards, and cooperation with the Type-Register, to computerized Floras and creation of major data centres. The advisory Committee for Systematic Resources in Botany identified 105 out of the 1127 Institutions as National Resource Collections, i.e. those considered vital for continued science support; they hold about 44 million specimens, over half of which are in just eight institutions. The total annual support needed was calculated at over 5.5 million dollars "for the continued well-being of the collections as research and information sources, not linked to the much greater costs of actual research and educational activities". Since these US herbaria represent about one-sixth of the world's total, the costs of simply maintaining the world's main herbaria, must be of the order of 30–35 million dollars per annum or in today's equivalent, 45–50 million dollars. The figures for zoological collections is at least of the same order, if not substantially greater, because of the more expensive methods of preservation used, so that it would seem that the global figure for maintaining the world's systematic collections is probably well over a 100 million dollars annually. Clearly, it is important that these costs should be kept under control by whatever means possible including of course the use of EDP. To the extent that taxonomy involves performing repeatedly the same functions in precisely the same manner (or functions that *could* be performed in the same manner), automation by EDP of certain procedures is both possible and desirable (cf. Shetler, 1974).

We must, of course, bear in mind Shetler's myths, and in particular the danger of databanking encouraging what he calls a curatorial syndrome:

> which in its worst form can reduce curators from scientists to registrars and book-keepers who label and record with consummate precision because they are more concerned about being able to explain what they have than why they have it.

On the other hand, it would often advance the cause of science if curators could indicate more readily what their herbarium contains, since why it is there is often a historical accident of no actual scientific importance!

A NEW FACE FOR TAXONOMY?

It is very important that we do not look upon EDP in taxonomy as no more than a means of performing existing procedures more rapidly or more efficiently. We should ask ourselves whether these procedures are any longer suitable for today's circumstances. Taxonomists will have to reconsider very carefully not only their procedures in capturing data (specimens, literature, etc. and handling them) but also the ways in which they may be presented to the public; not just the style of Floras and other hard publications, which are often outdated and reflect the ethic of another age (Heywood, 1984; Frodin, 1984), but whether they should be retained in a data bank and called up on screens (VDUs) as needed and in whatever combinations are needed. Perhaps we have become too obsessed with academic scholarship in taxonomy, and with the learned volumes representing authority.

This raises the question of new methods and procedures for users of taxonomy. It also raises interesting questions about the ways in which data can be controlled and about responsibility for machine-generated "decisions", i.e. the problem of authority. This has been posed by McNeill (1973) in the following terms:

> The present system of data publication in scientific journals or the production of a diversity of Floras for a single area ensures for the most part that all well-reasoned data presentations can be communicated, and the user can compare these and make his own judgements. Such a facility is easily lost in a databank which carries out any data synthesis as a taxon-based one does by its very nature. It is usually easy to update or "correct" stored data but taxonomists at least know that new does not necessarily mean better, and the method of control of revisions in a national or international data bank could have far-reaching effects on the development and utilisation of that whole field of knowledge.

NUMERICAL TAXONOMY TODAY

Most of the applications of computers to taxonomy in the past 15–20 years have been concerned with numerical classification, involving clustering methods and resemblance measures, and methods for revealing taxonomic structure. The field is commonly referred to today as numerical phenetics and has been well reviewed by Sneath (1976) and others. There are several text books available including a second version of Sokal and Sneath's classic 1963 sourcework (Sneath and Sokal, 1973). An excellent recent concise text has been provided by Dunn and Everitt (1982). Some kind of numerical phenetics is now a standard procedure in much taxonomic revisionary work, especially in universities. Frequently, an array of different phenograms is offered and compared with conventionally produced taxonomic schemes. There is, I sense, a certain feeling that numerical phenetics has not lived up to its promise of providing taxonomists with unambiguous and respectable methods for constructing classifications: this is, I suspect, largely a reflection of a misguided belief that there is one "true" or "correct" classification, while numerical phenetics will offer an array of different classifications according to the resemblance coefficients and methods of cluster analysis used. Another attitude is that numerically produced classifications are often so concordant with conventionally produced ones that they are not worth the additional effort involved. This agreement could be interpreted either as indicating that previous classifications are essentially "correct" or that it has not been possible to eliminate subjectivity in the numerical techniques so that one is reflecting neural classifications already published or suspected.

On the other hand, disagreements with conventionally produced classifications do occur and, while some of these may be the result of applying inappropriate numerical techniques (Sneath 1976), they may well represent the greater ability of numerical phenetics to resolve complex patterns shown by large amounts of data. This is especially true in such groups as microorganisms where there is a very limited range of morphological features available for classificatory purposes.

It can be argued that one of the greatest benefits to taxonomy derived from numerical methods has been the attention that it has caused to be focused on many of the basic procedures such as the concept and handling of characters, weighting, homology, etc.

Homology remains one of the weakest areas in phenetics (Sneath, 1976) and has been the subject of further attention recently by Hennigian cladists. Since the introduction of phenetics it is noticeable that many taxonomic papers, especially revisions, are much more explicit than formerly on procedures, specifying the database, character coding, etc.

I fully concur with Sneath's view (1976) that many of the unsolved problems of numerical phenetics stem from an inadequate formulation of taxonomic theory – in other words "numerical techniques cannot be developed until taxonomists can say what it is they wish to measure". It is no good asking for techniques that will provide the best possible classification until one is prepared to say how one will measure the goodness of the classification.

In the past few years, there has been an increasing emphasis on numerical cladistics (Estabrook, 1972), sometimes applied to biochemical data such as the cytochrome C and plastocyanin work of Boulter (1973) which caused a flurry of controversy when first published, although the original claims have now been largely moderated.

NUMERICAL CLADISTICS AND THE ART OF PHYLOGENETIC CONSTRUCTION

A feature of recent developments in cladistics has been the shift of emphasis away from the reconstruction of evolutionary history of taxa to an emphasis on characters, with the cladograms no longer representing a formalized phylogenetic tree but the sequence by which uniquely derived characters emerged. The latest version, called transformed cladistics, is no longer even concerned necessarily with evolution but, to use the words of Patterson (1980) "It is about a simple and more basic matter, the pattern in nature". The pattern is dichotomous and the nested synapomorphies of a cladogram represent taxonomic structure". Sneath has suggested the term synapomorphogram for these diagrams. In terms of databases, there are two important points to make here:

(1) Transformed cladistics is considered by its exponents as a means of describing character-state distributions amongst organisms from which a classification can be derived, and is in some ways similar to phenograms and the derived "natural", i.e. overall similar classifications of pheneticists. As long ago as 1967, Colless claimed that Hennigian cladistics is an intuitive prototypical form of statistico-phenetic taxonomy.

(2) The clear distinction established some 20 years ago between phenetics and phylogenetics, the recognition of the different evolutionary components of phylogeny – patristic, cladistic and chronistic, and the precise use of terms such as classifications, characters character state, cladogram, etc. have been lost by the use of a confusing new set of meanings by the transformed cladists (cf. Sneath, 1982).

CONCLUSIONS

In a report to the National Research Council on The Importance and Needs of Systematics in Biology in 1953, one of the recommendations made was that taxonomists should note that they have the obligation of presenting the available knowledge in a manner acessible to their botanical colleagues working in other fields . . . and to others needing botanical information (Just, 1954). Thirty years later the message applies with even more force.

In December 1982 a conference was held in London which marked the end of Information Technology Year in Britain. It does not appear to have been celebrated in any special way by taxonomists. Although somewhat belated perhaps, we can consider this symposium to be our contribution and assess how far we have progressed in applying information technology to taxonomy and systematics.

REFERENCES

Anon. (1974a). Trends, priorities and needs in systematic and evolutionary biology. *Syst. Zool.* **23**, 416–439.

Anon. (1974b). "Systematic Botany Resources in America Part I. Survey and Preliminary Ranking". American Society of Plant Taxonomists. New York Botanic Garden, New York.

Anon. (1979). "Systematic Botany Resources in America Part II. The Costs of Services". American Society for Plant Taxonomists. New York Botanic Garden, New York.

Boulter, D. (1973). The use of comparative amino acid sequence data in evolutionary studies of higher plants. *In* "Progress in Phytochemistry", vol. 3 (L. Reinhold and Y. Liwschitz, eds), pp. 199–229. Interscience, London.

Brenan, J. P. M., Ross, R. and Williams, J. T. (1975). "Computers in Botanical Collections". Plenum, London.

Colless, D. (1967). The phylogenetic fallacy. *Syst. Zool.* **16**, 289–295.

Cullen, J. (1984). Libraries and Herbaria. *In* "Current Concepts in Plant Taxonomy" (V. H. Heywood and D. M. Moore, eds), pp. 25–38. Academic Press, London and Orlando.

Dunn, G. and Everitt, B. S. (1982). "An Introduction to Numerical Taxonomy". Cambridge University Press, Cambridge.

ESF (1977). "Taxonomy in Europe". [ESRC Review, no. 13.] European Science Foundation, Strasbourg.

Estabrook, G. F. (1972). Cladistic methodology: a discussion of the theoretical basis for the induction of evolutionary history. *Ann. Rev. Ecol. Syst.* **3**, 427–456.

Frodin D. G. (1984). "Guide to Standard Floras of the World". Cambridge University Press, Cambridge.

Gray, B. (1980). Popper and the 7th approximation: the problem of taxonomy. *Dialectica* **34**, 129–153.

Heywood, V. H. (1973). Ecological data in practical taxonomy. *In* "Taxonomy and Ecology" (V. H. Heywood, ed.), pp. 329–347. Academic Press, London and Orlando.

Heywood, V. H. (1974). Systematics – the stone of Sisyphus. *Biol. J. Linn. Soc.* 6, 169–178.
Heywood, V. H. (1980). The impact of Linnaeus on botanical taxonomy – past, present and future. *Veröff. Joachim Jungius-Ges. Wiss. Hamburg.* 43, 97–115.
Heywood, V. H. (1984). Designing Floras for the future. *In* "Current Concepts in Plant Taxonomy" (V. H. Heywood and D. M. Moore, Eds), pp. 397–410. Academic Press, London and Orlando.
Heywood, V. H. and Clark, R. B. (Eds) (1982). "Taxonomy in Europe – Final Report". European Science Foundation, Elsevier, North Holland.
Jacobs, M. (1980). Revolutions in plant description. *Misc. Papers Landbouwhogeschool Wageningen* 19, 155–181.
Just, T. (1954). Generic synopses and their role in modern botanical research. *Taxon* 3, 201–202.
Krauss, H. (1973). The information system design for the Flora North America Program. *Brittonia* 25, 119–134.
Lindroth, S. (1978). Linnaeus in his European context. *In* "Yearbook of the Swedish Linnaeus Society. Commemorative volume", pp. 9–29.
McNeill, J. (1973). Review of "Data Processing in Biology and Geology" (J. L. Cutbill, ed.).
Mello, J. F. (1974). Computer revolution in systematics. *Taxon* 23, 21–22.
Patterson, C. (1980). Cladistics. *Biologist, Lond.* 27, 234–240.
Raven, P. H. (1977). The systematics and evolution of higher plants. *In* "Changing Scenes in Natural Sciences, 1776–1976", pp. 59–83. Academy of Natural Sciences, Philadelphia [Spec. Publ. 12], Lancaster.
Shetler, S. G. (1971). Flora North America as an information system. *BioScience* 21, 524–532.
Shetler, S. G. (1973). Information systems and data banking. *In* "Vascular Plant Systematics" (A. E. Radford, W. C. Dickison, and C. R. Bell, Eds). University of North Carolina Press, Chapel Hill.
Shetler, S. G. (1974). Demythologizing biological data banking. *Taxon* 23, 71–100.
Sneath, P. H. A. (1976). Phenetic taxonomy at the species level and above. *Taxon* 25, 437–450.
Sneath, P. H. A. (1982). Review of G. Nelson and N. Platnick, "Cladistics and Biogeography: Cladistics and Vicariance". *Syst. Zool.* 31, 208–217.
Sneath, P. H. A. and Sokal, R. R. (1973). "Numerical Taxonomy: the Principles and Practice of Numerical Classification". Freeman, San Francisco.
Taylor, R. L. (1971). The Flora North America project. *BioScience* 21, 521–523.

2 | Information Services in Taxonomy

F. A. BISBY

Biology Department, The University, Southampton SO9 5NH, UK

Abstract: Rapid growth in the use of computer databases in taxonomy raises the prospect not only of using databases for tasks formerly done manually, but also of making new connections or performing new tasks. My particular aim in this paper is to discuss, albeit speculatively, ideas on how databases might allow changes to take place in the information system that is provided by taxonomists for use by society. The current information system in taxonomy consists of text descriptions structured in accordance with the classification hierarchy and published in journals or books such as Floras, Faunas or monographs. Users commonly begin at three entry positions and the system, when it is operated successfully, leads to one or more named taxa and attached descriptive data. Several basic aspects of this system remain obscure, however. Are the data required by the user of the same kind as the comparative descriptive data used in making the classification? Who are the actual users? Are there other potential users? Areas where one might look to innovation from experiments with databases are: (1) frequency of revision; (2) handling the shifting structure of alternative classifications and nomenclatural disagreement; (3) inadequacies in customer relations (taxonomists' own products are highly technical but they eschew the more widely acceptable popular-style products); (4) linking the classificatory data of taxonomists with other types of data required by users; and (5) the difficulties of using generalized products for specific tasks.

INTRODUCTION

Why does the community need taxonomists and, indeed, pay considerable sums from the public purse to employ them? The reason is to be found in the introductory chapters to most taxonomic

Systematics Association Special Volume No. 26, "Databases in Systematics", edited by R. Allkin and F. A. Bisby, 1984, pp. 17–33. Academic Press, London and Orlando.

DATABASES IN SYSTEMATICS
ISBN 0 12 053040 6

textbooks: "to systematize data for the use of other disciplines" (Blackwelder, 1967, p. 6). Despite these passing references, in practice remarkably little attention is given to studying these "serious responsibilities towards society" (Davis and Heywood, 1963, p. 2). The chapters analysing information services and public responsibilities are conspicuous by their absence in these textbooks.

As one of their roles, taxonomists work to provide an information service which is essential to all who deal with organisms. I call this the *Taxonomic Information Service*. It deserves much greater scrutiny if taxonomists are to meet their responsibilities. In the present context of the role of databases we must also characterize this service carefully so that we can design parts for it. One goal is the production of information handling systems that enable tasks to be done more efficiently. A further goal, however, is to improve the system itself by performing tasks or making interconnections that could not be done before.

THE TAXONOMIC INFORMATION SYSTEM

1. Retail Model

A simple model for the taxonomic information system is a retail trade with products manufactured by professionals behind the shop counter and marketed to customers amongst the public in front of the counter. The two boxes in Fig. 1 enclose the activities either side of the counter. The *taxonomic products* are published Floras, Faunas and monographs each normally composed of four interwoven constituents:
– a classification arranged in a hierarchical system of taxa,
– a nomenclature for the taxa,
– descriptions of the taxa,
– aids to identifying the taxa.
Of these the classification is clearly the cornerstone, despite a feeling amongst customers that it is names, descriptions and identifications that count.

Each taxonomic product, however excellent or self-contained it may be, is just one item in a continuing service. Such products compete for acceptance and use, have a varied useful life, and sooner or later are partly or wholly replaced. The need for revision and alteration of products arises from the constant accretion of data and the change in ideas on how the data should be analysed and presented in the products. Further changes come from the need to coordinate products so that a stable interlocking system of classification, nomenclature and description is gradually built up for all organisms.

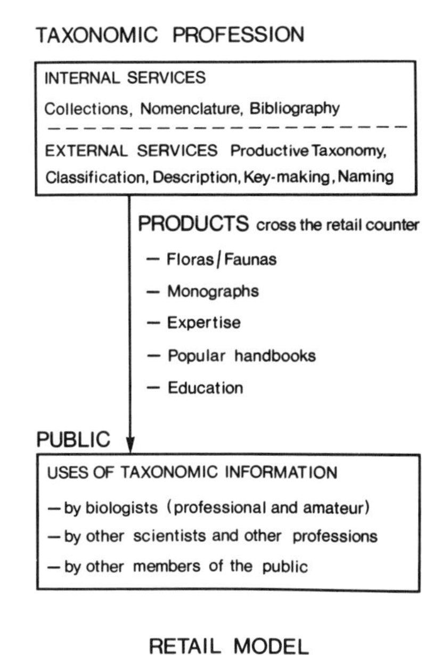

Fig. 1. A retail model for the taxonomic information service.

The *customers* referred to come from a wide range of disciplines both outside biology and within. Some obtain information directly from the taxonomic information system whilst others obtain it indirectly, as in conversation amongst gardeners or lectures at college, from people who have obtained it directly.

Lastly the *taxonomists* in this model are not just those working on taxonomic revisions. We must include those who work behind the scenes, nomenclaturalists and curators for instance, and those who work close to the counter interpreting technical products to a wider clientele, as for example in field guides or television programmes.

This Retail Model is a considerable oversimplification but it does suggest some useful perspectives. Should we distinguish more carefully between technical publications intended for use within the profession (which do not cross the counter) and products intended for use by non-taxonomists (which do cross the counter)? Should we put more thought into how technical taxonomic information is reprocessed and packaged for popular handbooks and educational materials? The model's shortcomings are evident when we think of the feedback in which data provided by other types of biologist are used by a taxonomist, or the "expert system" in which customers obtain information directly from an expert.

2. *Diffusion Model*

My Diffusion Model treats the taxonomic profession as something that funnels and processes certain kinds of information for dissemination in the community. Materials and data diffuse in, as at the top of the funnel in Fig. 2, and taxonomic products diffuse out through several zones depicted beneath the funnel. In the funnel, i.e. within the profession, are found those providing services for use by the profession and those doing primary taxonomic work, studying fresh

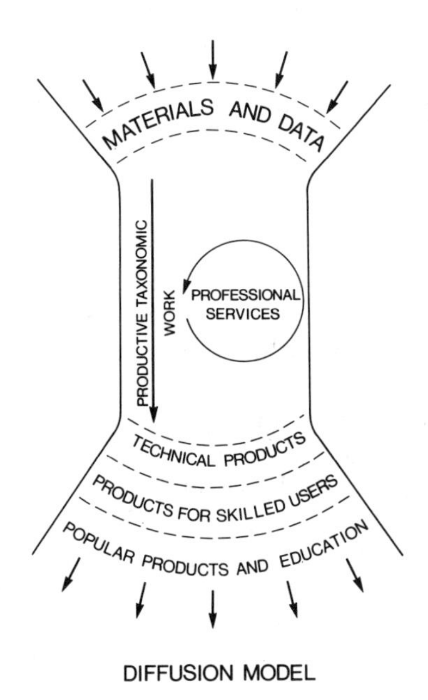

Fig. 2. A diffusion model for the taxonomic information service.

data as it comes in and making revisions that lead eventually to the products that diffuse out.

In the model the taxonomic products are not thought of so much as physical books as the intellectual information they contain, the classification, names, descriptions and identification aids. These products can then be translated in the diffusion process going from the technical descriptions, synonymies and keys of a monograph, through the "skilled-person's" descriptions, accurate illustrations and varied identification aids of field handbooks, skilled amateurs, and university courses, to the coloured illustrations, crude descriptions and common names used in elementary education.

In many ways this diffusion model is superior to the retail model. It reflects the difficulty in saying where taxonomy starts or where it finishes. It shows the productive and the housekeeping streams whose separation or integration causes such difficulty in taxonomic institutions. What it lacks is the pressures for self-discipline in the retail model where products for the public are clearly distinguished from those for the profession.

SOME BASIC QUESTIONS

Just as the taxonomic information system as a whole has been neglected, so has the area of customer relations. From what angles do customers approach the system for information? What kind of information do they want? And anyway who are the customers? A view held by many biologists is that it is identification which is of fundamental importance: "putting a name to it" is the purpose. At face value this is illogical. "Why not call it species A and not bother with identification" exclaims the ecology beginner faced with his first *Carex* or his first solitary bee. One aspect of identification does seem illogical: some humans do like to put a name to things even if the things are investigated no further. The other is that when talking of identification, people take for granted that the name acts as a tag for communication and that the communication in turn provides access to descriptive information.

1. Access to Taxonomic Information

My impression (for there are no statistical surveys that I know of) is that most customers try to gain access to the taxonomic information system at three entry-points – through names, through identification, or through descriptive data.

Imagine that a student hears of interesting properties of a named plant in a lecture, as in "Many Solanaceae have hallucinogenic properties". Either the name is familiar to him and conjures up his recollected information on the taxon named ("tomato-like plants with juicy looking berries") or he must go to the library and follow possibly several leads on the Solanaceae until he reaches a description. Perhaps a colour illustration, a formal botanical description, or a layman's description in a handbook meets the information, style and detail of his enquiry. The name has given him access and the information obtained is descriptive.

Or a customer enters at the identification point: "This plant has come up in my garden: what is it?" Producing the name often doesn't satisfy this demand. The customer really meant to add ". . . and where does it come from, is it poisonous or how tall will it grow?" Depending on his level of botanical knowledge this customer may or may not succeed in identifying the plant using a key, or he flips through the illustrations of a field guide or he gets help at a local museum. However the identification is made, he reaches first a name and then from this finds out more about the plant. Identification of a specimen provided the entrance to the system, but as before the target information was essentially descriptive.

Lastly, comes the difficult customer who wishes to retrieve either a name or descriptive data from a descriptive starting-point. A friend wants to know the name of a "good" evergreen shrub that will grow near the sea; a student wishes to know "more" about high forest trees in Amazonia that have seeds over 3 cm in length and seedlings that reach 6 m before the leaves open; a colleague wants to know what plants have "interesting" extra-floral nectaries. Even at their simplest (that is ignoring the words "good", "more" and "interesting"), these three requests are not catered for in the present system. Even if it were possible, searching *Flora Europaea* or *Flora Brasiliensis* from cover to cover would hardly yield authoritative answers. In practice, such questions are only answered, and then often inadequately, by asking botanists with the appropriate breadth of knowledge and by making informed extrapolations. For example, if tropical species of *Vicia* yield blood-group specific lectins then let us try British species of *Vicia* and see whether they do so too. The words "good", "more" and "interesting" imply, as with the identification enquiries, that when the names of appropriate taxa are located, further descriptive information will be requested. The target information is both a name and further description: the entry-point was retrieval based on descriptive characteristics.

What is interesting is that all three kinds of access involve names and descriptions: whatever the method of access, these constitute the target information.

2. Relation between Target Data and Classificatory Data

Our current taxonomic products and our designs for descriptive databases should surely be affected by how much overlap there is between the classificatory data used by taxonomists and the target data required by a customer. If all data are of interest both to taxonomists and to customers, like chromosome numbers, then clearly we might make descriptions which both characterise the taxa and meet the customers demands. There are, however, types of data used in classification that are likely to be of little interest to most customers, for instance the micromorphology of pollen ornamentation; and, conversely, types of data of general interest to customers that may be of little taxonomic value, for instance the size of a plant and its time of flowering. In such cases we could provide two descriptions, one giving the classificatory diagnosis, and one giving the customer's description.

The problem is more complex than this. Not only is there a variable degree of overlap between the classificatory data and the customer's target data, but where the overlap occurs there is variation in how best to express the data. The taxonomist, pigment chemist and field naturalist interested in the flower colours of gentians may yet demand quite different expressions of the same data.

This difficulty is met in traditional taxonomic products largely by providing a single generalized description which is a compromise. Occasionally the classificatory or the identificatory data are picked out, either in a special diagnosis or in italics as in Clapham *et al.* (1952). The alternative is to narrow the range of audience intended for the product. This happens in many popular books which concentrate on a certain kind of customer's data and omit classificatory data, perhaps just assigning species to families, or entering phrases of the type "technically Botanists place these two plants in different families". Conversely, there are publications that contain only technically expressed diagnoses; the *The Genera of Flowering Plants* of Hutchinson (1964) is one. Note that these do not count, at least in my Retail Model, as taxonomic products because they are not intended to cross the counter into the customer's domain.

3. Customers

Who are the customers for the taxonomic information service? I know of no wide-ranging survey. Indeed the ABRC Report (1979) and the ESF Reports (ESF, 1977; Heywood and Clark, 1982), are silent on the matter. The NERC Taxonomy Report (NERC, 1976) restricts itself to demands from United Kingdom workers in ecological research.

For the angiosperms, with which I am primarily concerned, there is clearly an enormous demand for descriptive gross morphology from field naturalists and from gardeners. Our bookshops are crowded with popular-level and skilled-level volumes competing in a very large trade, and in one year *The Concise Flora of the British Isles* (Keble Martin, 1965) was the best selling hardbacked book in the United Kingdom. Amongst professional customers there is again an assumption that field workers predominate: they are a large class of people and many of them oscillate awkwardly between using technical works such as *Flora Europaea* (Tutin *et al.* 1964–80), and popular works. Similarly, it is stated that Floras sell in larger quantities than monographs, although even here the numbers are quite small for "professional" Floras (Heywood, 1984). Little is said, however, about the very wide range of other users. Chief amongst these must surely be the information demands of either other biologists who are not taxonomists (the physiologists, biochemists, breeders, and husbandrymen of all kinds), or of the professions and businesses that use organisms (the chemists, the medical profession, the pharmaceutical industry, environmental agencies, agricultural industries, civil engineers, lawyers, architects, landscape designers, and so on). But I am guessing. Whilst a few institutions do keep records of requests for expert advice (Cullen, 1984), I believe there is a need both for a wide-ranging study to identify and quantify a profile of customers, and for a study on how their demands are met in the diffusion of taxonomic information, as in my Diffusion Model. I suspect that taxonomists may not do enough to meet the demands of the "other" professions, particularly where these demands range over several data types outside gross morphology, where monographic rather than floristic coordination is appropriate, and where professionals are used to rapid, efficient information services in other disciplines.

AUTOMATED TAXONOMIC INFORMATION SYSTEMS

In an earlier paper (Bisby, 1984) I review the slow development of taxonomic information systems up to 1982. The ideas of integrated taxonomic information systems of people like Rogers (see Estabrook and Brill, 1969) and Morse (1974) were ahead of their times, or indeed, ahead of the practicalities of taxonomic information management. The earliest projects either made valuable but limited progress with using database files for housekeeping activities (e.g. Flora de Veracruz), or, as in the case of Flora North America, foundered in the design stage. There is now a renewed interest in taxonomic information systems: severe economies are pressing in most institutions, information technology continues its explosive expansion with an equally dramatic fall in costs, and, perhaps most importantly, expertise in database management and database advances in software are building up. Now is the time to assess not only the practicalities of using databases for tasks formerly done by hand, but also to look ahead at the idealized systems which might allow new connections and the performance of tasks not previously done. If these new conconections or new tasks were to allow improvements to the whole taxonomic information system and, hence the service to the public, they would be of considerable value.

1. Types of Taxonomic Database

There are very many areas in the taxonomic profession which involve tasks of information management, collating, storing, checking, sorting, cataloguing, listing and retrieving data and, hence, where computer databases will without doubt prove useful, if not revolutionary. The principal applications have so far been in curatorial, biogeographic, bibliographic and conservation areas with somewhat less activity in nomenclatural and descriptive work (Sarasan *et al.*, 1983, Bisby, 1984).

I believe it is important that biological descriptions be included with nomenclatural data in designs for an integrated taxonomic database (Bisby, 1984), despite the special difficulties they present

(Allkin, this volume). As already discussed, the target for people using the current taxonomic information system is nearly always both a name and associated descriptive data. It is rare for people to want just a name on its own. Indeed, the name is a tag which only becomes of value when it is used to access further information. One of our goals then must be a descriptive database in which data, for instance, on morphology, chromosomes and chemistry is structured taxonomically and made available as an information service to the customers of taxonomy. The Vicieae Database Project reported elsewhere (Bisby *et al.*, 1983; Adey *et al.*, Chapter 15, this volume) is one experimental database of this kind.

2. *The Descriptive Database*

Because of the difficulties in handling biological descriptions, various suggestions have been made that descriptive databases should be stored either in textual or in codified form. As descriptions are usually published in textual form with at least simple sentence structure (albeit usually without verbs), and with a certain amount of subjectively determined phraseology, text formats such as those used in word processing and text editors would provide one way of handling descriptions. Such a textual form would be in contrast to the codified forms that have been used for some time in numerical taxonomy and in computer-produced identification aids.

There are, however, many advantages in using a codified form (Bisby, 1984). Perhaps the most significant is that with a codified database there is the prospect of using the data for several quite different purposes: retrieval of data subsets for direct output, description generation, identification aids, phenetic classification and cladistic analysis. The scheme illustrated in Fig. 3 has the codified descriptive database and character-based retrieval capability upstream of the five applications. If the retrieval of data subsets is followed by passing the data to the five applications, then a vast number of different products could be produced. Identification keys for the plants of different countries, such as the *Vicia* species of Spain, or classifications using different types of data, such as a classification of *Lathyrus* species on seed characteristics, would be amongst the possibilities. The very extensive labour of designing and constructing the codified database would then be amortized over the large number of uses to which it would be put. Notice too that the retrieval and four of the five applications areas are impossible if we switch to a textual database.

So far no system can perform the whole of the scheme adequately. The EXIR program linked with EXIRPOST and several application programs can lead retrieval data into all five applications (see Adey *et al.*, Chapter 15, this volume), but the flat-file data structure

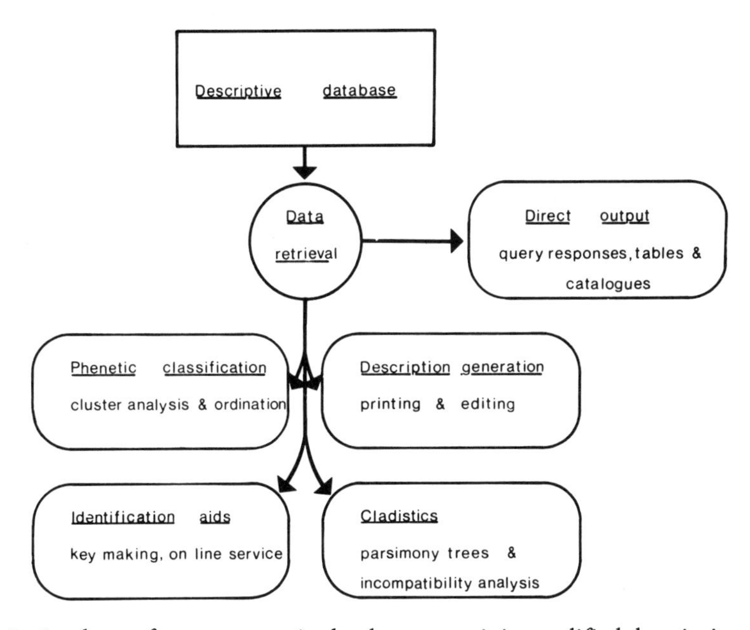

Fig. 3. A scheme for a taxonomic database containing codified descriptions and for its use in a variety of taxonomic activities.

required puts a severe restriction on the data that can be handled (see Brill and Estabrook, Chapter 5, this volume). Pankhurst's suite of identification programs (Pankhurst, 1975–79) and Dallwitz's CONFOR program (Dallwitz, Chapter 23, this volume) are both able to handle a wider range of data and to communicate the data with description generation and identification aid application programmes but neither performs character-based retrievals upstream of the applications.

3. Areas of Difficulty

The principal difficulty in implementing my scheme effectively is the need for a data structure that meets the very different demands of:

(1) including all the complexities of a taxonomic description, (2) making the whole, or at least the codified part retrievable by character-states, and (3) meeting the demands of the five application areas. Taxon variability and character dependence provide the greatest challenge.

Dallwitz's DELTA format (Chapter 23, this volume) allows the user to make textual comments at various points in a description. This overcomes one part of the taxon variability problem. Unique conditions or subjectively defined qualifiers such as "mostly" can be kept separately in these comments which can be retrieved complete when appropriate. The other aspect of variability, two or more states of one character occurring in the same taxon, gives difficulty when it comes to data retrieval. How, for example, can we arrange for a retrieval of species with hirsute pods to specify which species have all pods hirsute and which have only some pods hirsute? The system devised by Macfarlane and Bisby (see Adey *et al.*, Chapter 15, this volume) provides an awkward solution which could operate smoothly if incorporated into EXIR/TAXIR as a facility such as the multiple descriptor proposed by Brill and Estabrook (Chapter 5, this volume).

Other areas of difficulty such as character dependence, character formulation, character selection, and database structure are discussed in Allkin and White (1982), Bisby (1984) and Allkin (Chapter 22, this volume).

AREAS FOR DEVELOPMENT

I have outlined above my view of the present taxonomic information system and my speculative suggestions for how an automated taxonomic database might be assigned to support or even enhance it. I should now like to add to these generalized views a number of positive practical steps which I should like to see as areas for experimentation and development in the near future.

1. Revision Frequency

A well-managed computer database can be revised frequently to take account of expansion, contraction or correction of the data. A floristic or monographic database that was frequently revised would provide several advantages over the present system in which taxonomic products are rarely revised and only occasionally replaced.

It would be of some limited use even before it was complete. Monograph and Flora writers do of course make revisions to their private notes after the work has been published. Whereas the revised notes are now rarely published, they could be incorporated in the database so that anyone consulting it would be consulting the most up-to-date version. If the database incorporated "housekeeping records", say on bibliography or materials, then the work involved in these records could effectively be passed on to the next generation of taxonomists, an innovation that would save vast amounts of time spent rewriting literature and on herbarium searches.

What we do not know yet is, what appetite taxonomist's customers have for frequently revised products. Could the satisfaction of seeing the latest entries be muted by too frequent a change of classification or names? Clearly this advance would be of principal importance in the best known and most used groups or floras where data, ideas and theories change rapidly.

2. The Shifting Taxonomic Structure

For many users the shifting taxonomic structure makes tracing information in the present literature very difficult even when the information is there. How often has each of us wasted time or failed to locate our target data when alternative classifications (*Cytisus scoparius*, the common broom, being entered as *Sarothamnus scoparius* in many books) or nomenclatural changes (*Lygos monosperma* becoming *Retama monosperma* once the name *Retama* is conserved) have led us astray. There are many different ways in which alternative classifications and automatic synomymy tracking might be built into a database so that wc could tracc the description of a taxon directly from a synomym or compare the members of a genus in either X's or Y's classification.

3. Customer Relations

The fresh introduction of databases and modern communications technology provides an unprecedented opportunity to design and experiment with taxonomic information services of many styles and for the full range of potential customers. I should like to see a whole range of experiments in providing monographic information to applied biologists working on given groups, for instance legumes or aphids. I should also like to see the taxonomic profession take the provision of popular field-guides and educational aids more seriously, and couple this with experiments particularly involving the visual media – film, video, cable TV, viewdata and so on. In both my Retail Model and my Diffusion Model one sees that these popular works must be thought of as part of the taxonomic information service for which we are responsible.

4. Linking Data Types

The current literature is not good at linking taxonomically arranged data of different types such as chemical, morphological and ecological data. The shifting taxonomic structure, varying levels of confidence in the authentication of materials, varied geographical or taxonomic frameworks all mitigate against obtaining a straightforward answer to many quite simple enquiries.

I found one such example by looking for data on secondary substances and chromosome numbers for the genus *Cytisus* and one of its species, *C. scoparius*, the widespread common gorse. I chose two chemical compendia (Harborne *et al.*, 1971; Darnley Gibbs, 1974) and two chromosome compendia (Moore, 1982; Fedorov, 1974). In one of these the appropriate data were listed under *Cytisus* and *C. scoparius*. In the other three, to cull all of the appropriate data, the reader needed to know that *Sarothamnus* is often included in *Cytisus*, and that *S. scoparius* is the same species as *C. scoparius*. The three also left the reader in doubt as to whether *Sarothamnus* data had been incorporated in generalizations about *Cytisus*. Yet this problem did not occur because of errors: each of the four books had the data listed in the appropriate place according to its own editorial policy.

The increasing potential size of databases will, I hope, mean that we can store different kinds of data side by side without too high a premium on space. The increasing sophistication of database management systems means that where these data types are stored side by side we shall increasingly be able to link them relatively easily in coordinated retrievals. For instance, we plan that in an advanced version of the Vicieae Database, a single interrogation will span several data types to answer the request "list flower colour and flavonoid pigments for those Vicieae found in woodland in France".

5. Difficulties with Generalized Products

Many of today's taxonomic products cover all the taxa in the whole area for which the taxonomic work was compiled. (The commonest exception to this is the availability of some Floras in parts.) However, there are very many occasions when such comprehensive treatment is a hindrance to effective use of its contents.

Consider two difficulties that I have encountered in teaching undergraduates on field courses. In contrast with using Clapham *et al.* (1952) in Britain, using *Flora Europaea* in Austria was unpopular with our students because of the much greater length of keys and subtlety of leads needed to separate one amongst so many species. This is not just a matter of doing more work of the same difficulty: the larger number of taxa in a generalized work makes it difficult for the authors to find clear contrasting leads for the initial leads of a key. A second difficulty arises now that we run the course at a place in Spain where there is a rich legume flora. We cannot afford enough copies of *Flora Europaea* Vol. 2 containing the Leguminosae to satisfy a class of 30. The first of these difficulties might be overcome if a large Flora could be broken down into parts for given areas: a slimmer volume for Austria or the Alps would have been easier to use. The second could be overcome if large Floras were available both in bound volumes and in fascicles: we would happily purchase a dozen copies of a Flora Europaea Leguminosae fascicle were it available.

The possibility of printing different subsets of the data, and possibly even of generating keys for these different subsets is surely an important goal for taxonomic databases. Indeed, for codified databases there is the option of printing in several languages if carefully translated character lists can be substituted for the same data.

CONCLUDING REMARKS

Because so many taxonomic activities are controlled by the limits of record-keeping labour, I believe that databases will make a significant impact on my more optimistic goal – enhancing the taxonomic information service as well as simply letting us do what we currently do, more easily.

ACKNOWLEDGEMENTS

I am indebted to David Rogers, Margaret Adey, Bob Allkin, Mike Freeston and Richard White for valuable discussions on many of the issues raised in this paper.

REFERENCES

ABRC. (1979). "Taxonomy in Britain". HMSO, London.

Allkin, R. and White, R. J. (1982). "Design criteria for a computer program to facilitate the acquisition, storage, retrieval and reformatting of biological descriptions". Southampton University Research Fund Papers.

Bisby, F. A. (1984). Automated taxonomic information systems. *In* "Current Concepts in Plant Taxonomy" (V. H. Heywood and D. M. Moore, Eds), pp. 301–322. Academic Press, London and Orlando.

Bisby, F. A., White, R. J., Macfarlane, T. D. and Babac, M. T. (1983). The Vicieae Database Project: experimental uses of a monographic taxonomic database for species of Vetch and Pea. *In* "Numerical Taxonomy" (J. Felsenstein, ed.), pp. 625–629. [NATO. Advanced Study Institute Series, Vol. G1.] Springer-Verlag, Berlin, Heidelberg and New York.

Blackwelder, R. E. (1967). "Taxonomy: a text and reference book". John Wiley & Sons, New York.

Clapham, A. R., Tutin, T. G. and Warburg, E. F. (1952). "Flora of the British Isles". Cambridge University Press, Cambridge.

Cullen, J. (1984). Libraries and Herbaria, *In* "Current Concepts in Plant Taxonomy" (V.H. Heywood and D. M. Moore, Eds), pp. 25–38. Academic Press, London and Orlando.

Darnley Gibbs, R. (1974). "Chemotaxonomy of Flowering Plants", 4 vols. McGill-Queen's University Press, Montreal.

Davis, P. H. and Heywood, V. H. (1963). "Principles of Angiosperm Taxonomy". Oliver & Boyd, Edinburgh.

ESF (1977). "Taxonomy in Europe". [ESRC Review, No. 13] European Science Foundation, Strasbourg.

Estabrook, E. F. and Brill, R. C. (1969). The theory of the TAXIR accessioner. *Math. Biosciences* 5, 327–340.

Federov, H. (1974). "Chromosome Numbers of Flowering Plants". Koeltz, Koenigstein.

Harborne, J. B., Boulter, D. and Turner, B. L. (eds) (1971). "Chemotaxonomy of the Leguminosae". Academic Press, London and Orlando.

Heywood, V. H. and Clark, R. B. (1982) (eds). "Taxonomy in Europe – Final Report". Elsevier, North Holland.

Heywood, V. H. (1984). Designing Floras for the future. *In* "Current Concepts in Plant Taxonomy" (V. H. Heywood and D. M. Moore, eds), pp. 397–410. Academic Press, London and Orlando.

Hutchinson, J. (1964). "The Genera of Flowering Plants", 2 vols. Oxford University Press, Oxford.

Keble Martin, W. (1965). "The Concise British Flora in Colour". Ebury Press, London.

Moore, D. M. (1982). "Flora Europaea check-list and chromosome index". Cambridge University Press, Cambridge.

Morse, L. E. (1974). Computer programs for specimen-identification, key construction and description-printing using taxonomic data matrices. *Publs. Mich. St. Univ. Mus., biol. ser.* 5, 1–128.

NERC (1976). "The Role of Taxonomy in Ecological Research". National Environmental Research Council [Publ. Series B no. 14.] London.

Pankhurst, R. J. (1975–79). Internal documents 1–7 from R. J. Pankhurst, British Museum (Natural History), London.

Sarasan, L. and Neuner, A. M. (eds) (1983). "Museum collections and computers". Association of Systematics Collections, Lawrence.

Tutin, T. G., Heywood, V. H., Burges, N. A., Valentine, D. H., Walters, S. M. and Webb, D. A. (eds) (1964–80). "Flora Europaea", 5 vols. Cambridge University Press, Cambridge.

3 Current Database Design — the User's View

D. W. BARRON

Computer Studies Department, The University,
Southampton SO9 5NH, UK

Abstract: "Database" has become one of the buzzwords of the computer industry. Its practitioners have surrounded it with a great mystique, largely derived from the invention of a new vocabulary, and it is regarded as very much a matter for the specialist. In this paper we aim to dispel the mystique, and to present the three philosophies – hierarchic, network and relational – in terms that can be understood by the ordinary user. Particular attention is paid to the relational approach, which seems to offer most promise for the development of user-oriented systems.

INTRODUCTION

Computer professionals have been likened to a priesthood, maintaining an aura of mystery about the computer so that none might come to it save by them. Mainframe software systems are indeed complex, and do require the services of experts if they are to be used successfully as aids to scientific research. One of the major benefits of the rapid spread of personal computers is their demonstration that software does not need to be complicated, and that an averagely intelligent scientist can use a computer productively in research without the aid of systems analysts, systems programmers and database administrators. The area of databases is perhaps the most recent one to be shrouded in mystery, as the "professionals" have developed

Systematics Association Special Volume No. 26, "Databases in Systematics", edited by R. Allkin and F. A. Bisby, 1984, pp. 35–42. Academic Press, London and Orlando.

DATABASES IN SYSTEMATICS
ISBN 0 12 053040 6

their own vocabulary to make it seem more difficult (and perhaps to justify their existence).

What then is a database? Fundamentally, it is nothing more than a collection of data relevant to some organization or enterprise. The essential characteristic is that the data are organized in such a way that they can be *shared* by several programs: they are not tied to a single program in a permanent symbiosis. This in turn leads to a different philosophy of programming, in which we regard the data as the central and important resource, which has an existence independent of programs that use or change it. Advantages of this approach are:

(1) There is no duplication; each data item appears exactly once in the database, no matter how many programs use it. Thus the database can be kept consistent, and problems are avoided that arise if one program updates a data item, whilst another program, with its own copy of that item, does not update it.

(2) It is possible for different programs to have different "views" of the data, i.e. some programs may be restricted to a subset of the total database.

(3) It is possible to provide device independence, so that programs do not need to be aware of the physical charcteristics of the storage medium. This makes it possible, for example, to change the data storage from floppy discs to a hard "Winchester" disc without any changes to the programs that access the database.

THE TRADITIONAL VIEW OF DATA

The traditional view of data is that it consists of *files* of *records* which are made up of *fields*, e.g. a fish wholesaler might maintain files of suppliers, items and shipments with a structure shown in Fig. 1.

Database Architecture

There are three basic approaches to the construction of a database – hierarchical, network and relational. We shall look at each in turn, using our hypothetical fish merchant to illustrate the differences.

1. The Hierarchical Approach

The hierarchical approach, typified by the IMS system provided on IBM mainframes, represents data in a simple tree structure. The database designer has to decide on the hierarchy: for our example he might choose to put items above suppliers. We would then get

Suppliers file

supplier_no	name	address
S1	Jones	Grimsby
S2	Smith	Grimsby
S3	Brown	Yarmouth

Items file

item_no	description	weight	port
I1	dabs	8 oz	Grimsby
I2	plaice	16 oz	Grimsby
I3	plaice	2 lb	Grimsby
I4	bloaters	large	Yarmouth

Shipments file

supplier_no	item_no	quantity
S1	I1	50
S1	I3	100
S2	I1	40
S2	I2	100
S3	I4	400
S4	I3	100

Fig. 1. Files for a fish wholesaler.

a structure as illustrated in Fig. 2. The user sees a number of trees, one for each item. The dependent records in the trees represent shipments, and so consist of a supplier record with a quantity appended. It will be observed that this is a natural extension of the file concept, but it is not a natural representation of the data. Some queries are easily processed, e.g. "find suppliers of dabs" merely requires a search of the tree under the item entry "dabs". However, the query "what fish does Jones supply" is much more difficult, since it involves searching every tree to see if "Jones" appears as a supplier. The apparent symmetry of the queries has been lost; this is not surprising since the tree structure is essentially assymmetric, with items superior to suppliers. In general, the hierarchic approach does not lend itself to data that involves many-to-many relationships.

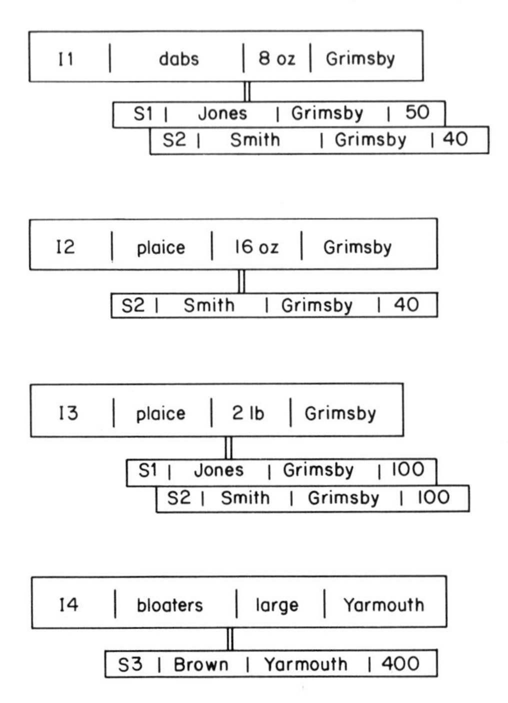

Fig. 2. Example of a hierarchical database.

2. *The Network Approach*

The network approach to databases was pioneered by CODASYL (the people who brought us COBOL), and is often called the CODASYL model. Typical of this kind of system is IDMS, available on many mainframes. In this approach the data is represented by *records* and *links*. Instead of the tree structure of the hierarchical model, we have a general network in which any particular occurrence of a record can have any number of superior records, as well as any number of lower-level dependent records. Returning to our fish wholesaler, there will be records for the items and suppliers: there is also a new kind or record called a *connector* to represent shipments. A shipment associates an item and a supplier: the connector records the quantity and involves two fields for chains. One of these is used to link together on a circular chain all shipments for a particular supplier, the second chain links the connectors for all shipments of a particular item. These structures are shown in Fig. 3. If we require to know "who supplies dabs" the record for that item is first found. The chain from this item will take us through a sequence of connectors, and from each connector we can follow the other chain which will lead (eventually) to a supplier record. Not surprisingly, this process has been described as "navigating the database".

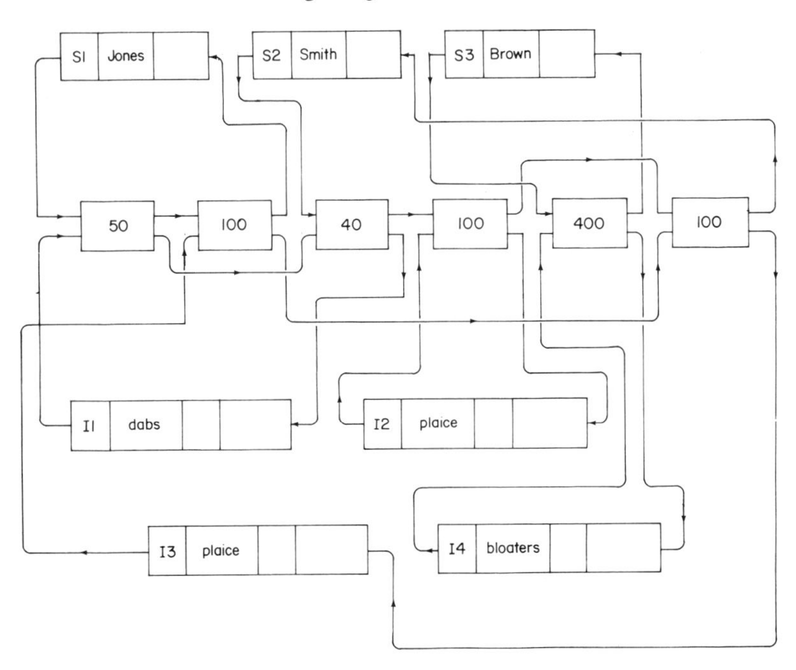

Fig. 3. Example of a network (CODASYL) database.

The CODASYL model has one major disadvantage: it requires a *program* to process any query. This type of database usually has an interface to a *data manipulation language* (DML) from COBOL: the DML appears as an extension to COBOL, and is used for the navigational programs required. Thus the procedure for using this kind of database is to decide what you want, tell the expert, and wait for him to produce the program. The CODASYL model can be criticized as being implementation oriented: the designer is always conscious of how material is stored, and what links are to be provided between records. One feels that the authors of the original CODAYSL proposals were intoxicated by their rediscovery of chained storage structures. (The same criticism of implementation-orientation applies to the hierarchical model as well, though to a lesser extent.)

3. The Relational Model

This is in fact the simplest and most user-oriented model, though its attractions were for a long time concealed by the fact that its inventors insisted on dressing it up in an abstruse mathematical formulation, and using unfamiliar mathematical terms for familiar notions.

The relational model is based on an idea so simple that it escaped many software designers who couldn't see beyond their CPU. Data is represented as *tables*, a familiar form that has been used by businessmen and scientists for untold years. The rows of the table link together related values, and different tables may be linked implicitly if they include values of a common item, e.g. a name. The fish merchants database is illustrated in relational form in Fig. 4. A table

Suppliers relation

S#	name	address
S1	Jones	Grimsby
S2	Smith	Grimsby
S3	Brown	Yarmouth

Items relation

I#	name	weight	port
I1	dabs	8 oz	Grimsby
I2	plaice	16 oz	Grimsby
I3	plaice	2 lb	Grimsby
I4	bloaters	large	Yarmouth

Shipments relation

S#	I#	quantity
1	1	50
1	3	100
2	1	40
2	2	100
3	4	400
2	3	100

Fig. 4. Example of a relational database.

is called a *relation*, a row is called a *tuple*, and a column is called a *domain*. Tables are manipulated using three basic operators:

- SELECT extract a horizontal subset from the relation (i.e. selected rows).
- PROJECT extract a vertical subset from the relation (i.e. delete certain columns) and remove duplicate rows from the subset so obtained.
- JOIN join tables where rows have common values in a given domain.

The use of these operators is best illustrated by some examples. "What shipments is supplier 3 making" is answered by

SELECT shipments WHERE S# = 3

which will select only those rows in the shipments relation that have supplier number (S#) equal to 3.

Usually, more than one relation will be involved. For example, suppose we have the query "what shipments is Jones making"? The shipment relation does not know about names, so we must use the supplier relation also, as follows.

SELECT suppliers WHERE NAME = "Jones"
JOIN WITH shipments OVER S#
PROJECT OVER item_name, quantity

The SELECT operation gives us a relation (in fact with only one row) containing the S# for Jones. JOIN compares this with shipments, and selects rows where the two relations have the same value for S#, giving an augmented relation containing the domains (columns) of both shipments and suppliers. Finally, the PROJECT selects the columns that we want to display as the result of the query.

The beauty of the relational model is that its way of representing data is intuitively "obvious" and natural to the user. The examples above have used an algebraic approach ("relational algebra") but it is also possible to frame queries in a "relational calculus" which is completely non-procedural, i.e. the user specifies *what* information he requires, not *how* to extract it, e.g. "item_name, quantity for supplier = 'Jones' ".

CONCLUSIONS

Despite the evident superiority of the relational approach to data, most of the databases in current use in the commercial world are based on the network approach. Why should this be so? One reason is that the CODASYL model blends nicely with COBOL; another is that CODASYL is largely dominated by the mainframe manufacturers, and it is natural that they should promote their own systems. A third reason is that the prototype relational systems were very slow: some searches were described as "taking weeks". One response to this has been the development of "backend" database engines, i.e. specialized machines which have the operations required for relational databases built into the instruction set. However, it has recently been found possible to implement relational databases of moderate size with reasonable efficiency on relatively small machines (e.g. CP/M systems). There is no doubt that the user's view of current database design, for small machines at least, is a relational one.

REFERENCES

Useful references of a tutorial nature are:

Date, C. J. (1981). "An Introduction to Database Systems", Addison Wesley, Menlo Park.

Martin, J. (1975). "Computer Data Base Organization", Prentice Hall, Englewood Cliffs.

Widerhold, G. (1977). "Database Design", McGraw Hill, New York.

4 | The Implementation of Databases on Small Computers

M. W. FREESTON

Computer Studies Department, The University, Southampton SO9 5NH, UK

Abstract: It is hard to keep informed of developments in two academic disciplines at the same time. It is clear from other papers contributed to this conference that some people manage to do it, but there must be those interested in taxonomic databases who cannot find the time to catch up with the computing aspects of the subject. I address myself to them, certain that any lack in their knowledge of Computer Science is more than matched by my ignorance of Taxonomy. I examine two areas: first, the hardware resources required for a viable database system, and in particular the extent to which it is now possible to consider the use of a microcomputer, rather than a mainframe; and secondly, the growing availability of commercial database software on microcomputers.

INTRODUCTION

I see my role in this paper as that of the Arab in the well-known cartoon in which a crusader is about to go into battle, while the Arab tries to persuade him to buy a machine-gun. The crusader, sword aloft, snarls "Don't bother be now – there's a war on."

Many of those involved in Systematics have used computers in sophisticated ways for a long time, but there are clearly many who could benefit substantially if they were to grasp the nettle and computerize their work. I use the phrase "grasp the nettle" advisedly. For those who have not interested themselves in computers, the prospects must be rather daunting.

Systematics Association Special Volume No. 26, "Databases in Systematics", edited by R. Allkin and F. A. Bisby, 1984, pp. 43–52. Academic Press, London and Orlando.

DATABASES IN SYSTEMATICS
ISBN 0 12 053040 6

There is also a frequent, though rarely expressed, fear that if a computer is used, control of the project will pass into the hands of a large and ever-growing band of computer gurus. This fear is not irrational; it has happened, often with most unfortunate results. Even those who have considered using a computer have often in the past been put off by the cost of the hardware, the extra staff and the programming effort required. But in the computer world, what was impractical last week may well be feasible, if not next week, at least within the next few months. The cost of computing power also continues to fall dramatically. So, if using a computer has previously been ruled out, it would be wise to review that decision at intervals of not more than a year.

There is a plethora of publications available on the bookstalls detailing recent advances in computing. For those who only have the time or inclination for an occasional glance through them, it is difficult to distinguish genuine advances from the advertiser's poetic licence. My objective is to draw your attention to some of these advances and to emphasize that quite large applications may now be implemented with computers which look ridiculously small, and cost relatively little.

DEVELOPMENTS IN COMPUTER HARDWARE

The rate of development of computer hardware is phenomenal, and shows no sign of slowing up. The significance of the development of microchips lies not in any advance in the design of computer architecture, but in their cost and reliability. The central processor of a computer of equivalent power which cost £10 000 ten years ago, costs less than £10 today. (Do not hope to buy a computer for £10. All the bits and pieces that go with the processor raise the minimum price for a professional computer to around £2500.) As for reliability, mean time-between-failures is now measured in years rather than hours.

The power of a computer can be judged approximately by the speed with which it can process data, and by its data storage capacity. The processing speed of the more advanced microcomputers is roughly equivalent to that of the minicomputers which came onto the market at the beginning of the nineteen-seventies. The reduction in the size and cost of central computer memory has had an equally dramatic effect. It is now possible to extend the memory of some microcomputers to several megabytes (millions of characters), at a cost of about £5000 per megabyte, representing a reduction in cost of at least a factor of ten, and a capacity which was previously only available on mainframe computers. (As a guide for those not used to quantifying data in terms of number of characters, the Bible contains approximately 5 megabytes.)

Equally important advances have been made in the technology of backing storage devices, required for the permanent storage of data. The development of the Winchester disc has been particularly valuable: because it is hermetically sealed, it does not require a dust-free atmosphere and, as a result of this and its overall construction, its reliability is extremely high. Mean time-between-failures is usually quoted well in excess of 10 000 hours of operation. Winchester discs are very compact, frequently occupying less volume than a nine-inch cube. Their data storage capacity, however, is continually increasing. Although discs of between 5 and 80 megabytes are the most common, 500-megabyte Winchesters are now being produced. A 5-megabyte disc at present costs about £1000. 80 megabytes might cost £10 000.

There is one obvious danger in the use of such backing-storage devices: when they do break, all the data stored on them will almost certainly be destroyed. The larger the capacity of the disc, the greater the disaster. It is therefore essential to provide protection by archiving to some removable storage medium, which might be an exchangeable disc, floppy disc or cartridge, or reel-to-reel magnetic tape. The choice depends on the quantity of data stored and the frequency with which the data are modified. In small systems, a popular combination is a Winchester disc of between 5 and 10 megabytes and one or two exchangeable floppy discs of between 400 kilobytes (thousands of characters) and 1 megabyte.

One other recent development in computer hardware is particularly striking. Computer terminals are beginning to appear on the market which have far more sophisticated display capabilities than the old 24-lines-of-80-characters type. They have a very high resolution of almost photographic quality in monochrome or colour. The display is essentially graphical, so that you can draw anything you like on the screen. Text can be displayed in any pitch and chosen from several founts. If none of those available are suitable, you can design your own. Displays in Arabic, for example, present no difficulty. Texts can thus be prepared precisely in the form required for printing. To produce a direct print-out of the same quality from the computer requires an expensive printer, but a cheap alternative is to photograph the screen.

This kind of terminal is made possible by devoting a microprocessor and as much as 1 megabyte of memory exclusively to support the display. It has only become economically viable because of the cheapness of chips. Nevertheless, such terminals are still expensive, but we may expect that they represent a standard for the near future, and that their cost will fall as dramatically as have nearly all the other elements of hardware.

DEVELOPMENTS IN SOFTWARE

By contrast with hardware, the development of software has been extremely slow. For example, the two programming languages which are most widely used throughout the world – COBOL and FORTRAN – have held that position for 30 of the 40 years, since electronic computers were invented.

It is not that there has been a lack of effort in the study of software. The slow growth reflects the fact that design and implementation of software is an arduous and labour-intensive activity. The Parkinson's Law of Software states that if there are more than two programmers in a team, doubling their number is likely to double the length of the project. One area of study which has however made significant practical advances in the last ten years is the development of database management systems (DBMS).

The principles of database management systems are described in Barron (Chapter 3, this volume). For a long time they were regarded as the exclusive province of large organizations with very large computers. Even now there is widespread ignorance of the subject, even within the computer industry. There are several reasons for this. Large organizations set up a DBMS because they recognize its potential value to coordinate their many activities. Because they are large organizations, they have a large computer. This has led people to think that a DBMS requires a large computer, and is only of value to large organizations. Large organizations can also afford to employ computer programmers who have a particular expertise in the use of a DBMS. There has not, therefore, been any great pressure to make the implementation of a DBMS easy to understand. Early DBMS were based on the hierarchical or network models, which, particularly the latter, are *not* easy to understand. The relational model is easy to understand, because it is based on the notion of tables, with which everyone is familiar. However, perhaps because it was a late starter, and perhaps because it was embraced too closely by the academic community, the relational DBMS has taken a long time to emerge into the commercial world.

Since 1980 many software packages have become available which claim the title of a relational DBMS. Hardly any of them satisfy the requirements to justify the title, but that doesn't matter too much. The general idea of a DBMS has taken hold, and once the principles are more widely understood and appreciated, demands for more complete systems will surely follow. Most importantly, DBMS have now been developed specifically for microcomputers, and designed to reach a much wider public. A detailed description of a number of such systems can be found in Kruglinski (1983).

GUIDELINES FOR THE EVALUATION OF DBMS

If at all possible, get the advice of an expert. However, since some are more expert than others, and since you are likely to know more about the nature of the application than anyone else, it is wise to be well informed of the criteria which should be used in judging a DBMS.

1. Cost

Prices vary from a few hundred to several thousand pounds, but the correlation between price and performance is weak. The most important point is that a DBMS provides a set of very powerful "software tools" which should make it possible to write programs for specific applications much more quickly than if starting from scratch. The cost of the DBMS must therefore be set against this saving in programming costs and time.

2. Portability

It is essential to maintain a stable interface between users and the DBMS. The DBMS should only be changed if it proves inadequate for its task. Since hardware is developing much faster than software, there is likely to be pressure to improve the efficiency of existing software by changing to a newer and faster computer. If the DBMS has been written in a programming language which is specific to one computer, this will not be possible unless an alternative version has been written for the new computer. This difficulty applies to many DBMS and other commercial software, particularly on microcomputers. Why do software authors limit the availability of their products in this way, when there are a number of programming languages (e.g. BASIC, FORTRAN and PASCAL) common to most computers? The main reason is that programs written in "low level" machine-specific languages are more efficient, and this has been a particularly important consideration for the first generation of microcomputers.

In recent years, however, the increasing power of computers and the advantages of portability have begun to outweigh considerations of efficiency. This trend is now apparent in the development of software for the second-generation 16-bit microprocessors, but it has not yet reached the microcomputer DBMS. However, alternative versions of most DBMS have been written for different machines. Because the number of microprocessors in widespread use is small, and is likely to remain so, this sledge-hammer approach to the portability problem may continue although it does not give true portability and delays the transfer of the DBMS to new processors. So, at present, there is little scope for selecting a DBMS which is portable, but this should be investigated when evaluating new systems.

3. Ease of Use

Relational systems are undoubtedly easier to use than network or hierarchical DBMS. It should be possible to create files and manipulate their contents in quite complicated ways without writing any additional programs. A good system should present the user with a clear, logical representation of the data items and the relationships between them. The internal organization should be completely hidden. Some systems, e.g. RAPPORT, require the user to state the maximum file size when the file is created and provide facilities to "tune" the access mechanisms to records in the file. Although the correct choices result in very fast access to records, it may not be possible to make a prior estimate of the maximum file size; and a clear understanding of the internal access mechanisms is asking too much of ordinary users. A system which sacrifices some speed for the sake of flexibility and conceptual simplicity is surely preferable. dBASE II is particularly good in this respect. On the other hand, unless experienced programmers are available, network systems should be avoided. Kruglinski (1983) describing "MDBS III" concludes: "It will take up to a month just to learn the basics."

4. *Data Sublanguage*

The power of a DBMS depends largely on the facilities provided to manipulate the data in the files. A set of commands must be available to add/amend/delete data items or records, and to combine or resolve them in various ways. This set of commands constitutes the data sublanguage of the DBMS. In network systems the commands are essentially *record-based*, i.e. the commands operate on one record at a time. In truly relational systems the commands operate on relations (or tables) as a whole, and the result of an operation is itself a relation, even if it contains only one record. For example, a table might contain a set of records describing attributes of various plants. An operation to extract from the table those records for which the flower colour is blue would in general yield several records, thus constituting a new table. The essential point is that in a relational system this operation would be expressed in a single command, whereas in a record-based system the operation would have to be expressed as an iterative sequence of commands, searching through the records for the next occurrence of a blue flower, returning with that record, and repeating the sequence until the end of the table is reached. The relational approach is more concise, and can be expressed in a form close to natural language, e.g. find all flowers with colour = blue.

It has been shown that there is a particular set of commands which can be combined to express any such query. A data sublanguage containing all these commands is said to be "relationally" complete. Unfortunately, as far as I am aware, there is no DBMS for microcomputers which approaches relational completeness, and the description of "Relational" DBMS is often applied quite indiscriminately to systems which simply represent their data in tabular form, and are otherwise record based.

A detailed discussion of Relational Data Sublanguages and the notion of completeness can be found in Date (1981) and Codd (1972).

5. Applications Programming

Unless a specific application of a DBMS is very simple, additional programs will have to be written to perform functions which are not provided in the generalized software. Some mechanism must therefore be available for incorporating these programs into the system. The usual approach is to write the programs in a "high-level" language such as FORTRAN or COBOL, but including data sublanguage commands within the program. Since these "embedded" commands are not part of the "host" programming language, there must be a special preprocessor which translates the data sublanguage commands into the host language. The complete program can then be compiled in the usual way. Preprocessors may be provided for several different programming languages, and it is important to establish which languages are supported by a particular DBMS. It is also important to bear in mind that so-called "standard" programming languages may be defined at several levels of standardization, e.g. FORTRAN and PASCAL, or no standard at all. The machine, the compilers and the DBMS must be compatible. MDBS provides interfaces with most widely used programming languages. RAPPORT provides an interface with FORTRAN.

dBASE II adopts a different approach. It provides its own programming language, which is a rather curious amalgam of other languages. It is very simple and easy to learn, but is barely adequate for use by professional programmers.

6. Multi-user Applications

The first generation of 8-bit microcomputers were designed as single-user machines, and did not contain the features necessary for true multi-user operation of a DBMS. With the arrival of 16-bit machines multi-user operation is much more practicable, and new versions of DBMS are being released offering this facility. The transition from single to multi-user is not as simple as it might seem. The problem is that both the computer operating system and the DBMS must be able to service the requests of several users without confusing their activities. This is most liable to occur when two or more users try to update a file at the same moment. To prevent confusion there must be some mechanism by which one user can lock a file or record against access by the others until the update is complete. This facility is essential for true multi-user operation of a DBMS.

7. Data Protection

With a personal computer the best data protection is achieved by taking away the key to the computer. In a multi-user environment password protection should be provided for files, programs and system access. It should also be possible to control whether an individual user has read or write access to a file. For the reasons given above, data protection has not until recently received the attention it now deserves in microcomputer DBMS. "dBASE II" for example, has no such facilities as yet.

8. Data Security

Data can be lost or corrupted due to user error, software error, hardware failure, physical damage or by a deliberate act, even if data protection measures are taken. The simplest protection against such eventualities is to take frequent back-up copies of all data files. This does not require any special database facilities. The frequency of back-up is decided by how much work you are prepared to face in re-entering lost transactions. It is also assumed that a written log is maintained of all the transactions since the last back-up.

In a single-user environment this may not be practical, and in a multi-user environment it almost certainly will not be. There is the additional possibility that an error may corrupt subsequent valid transactions, but go unnoticed for some time. Ideally, the DBMS should deal with these problems by maintaining its own internal record of all transactions and initiating its own recovery procedures when an error is detected. This is not straightforward, and no entirely satisfactory solution has yet been devised. Nevertheless, "RAPPORT" and "MDBS" do contain error recovery features.

CONCLUSION

Aspects of developments in computer hardware have been discussed to emphasize how powerful computers can be, even if they look insignificant; why there is no need to become involved with air-conditioned rooms, and hordes of engineers; the large quantities of data which can be stored in "small" computers; and how much you will get for your money. It is important to recognize that professional microcomputers are not toys, and that there is no longer any clear division between micro- and minicomputers.

In discussing the development of computer software, I have concentrated entirely on database management systems to underline that such systems are no longer the exclusive preserve of a few experts in large organizations. Powerful systems are available on very small computers. Although they cannot achieve their full potential if the user does not write additional programs, the attraction of relational systems in particular is that the user can use them profitably with a minimal previous knowledge of computers. By providing a "high-level" interface with the computer, immense savings in programming effort can be made. A system common to all users forces a high degree of coordination upon them and a project leader is thus able to maintain firm control. Since the software for a DBMS on a microcomputer can cost as little as £400, no one should begin even the smallest database project without considering the advantages of such a system.

Finally, if after careful consideration you cannot find a computer which is both cheap enough and powerful enough, don't despair – try again next year!

REFERENCES

Codd, E. F. (1972). Relational completeness of data base sublanguages. *In* "Data Base Systems" [Courant Computer Science Symposia Series, Vol. 6] (R. Rustin, ed.), pp. 65–98. Prentice-Hall, Englewood Cliffs.

Date, C. J. (1981). "An Introduction to Database Systems". 3rd edition. Addison-Wesley, Reading, Mass.

Kruglinski, D. (1983). "Data Base Management Systems: A Guide to Microcomputer Software". Osborne/McGraw-Hill, Berkeley, California.

5 Management of Almost Flat Files in Systematic Biology Using TAXIR

R. C. BRILL

University of Michigan Computing Center, Ann Arbor, Michigan 48109, USA

and

G. F. ESTABROOK

University of Michigan Herbarium, Ann Arbor, Michigan 48109, USA

Abstract: TAXIR, a database management system written in the late 1960s for large mainframe computers has since undergone many improvements, and has been converted to run on a variety of machines and in a number of operating environments. The flat file structure (a single relation) of TAXIR databanks enables efficient storage structures and very rapid retrieval, but is not well suited to data structures that cannot be construed as flat files. Here we present some methods to flatten database files that are not quite flat for use with present versions of TAXIR. Finally, we sketch a design for increasing the powers of TAXIR to accept directly non-flat structures, which we call prejoins, which when implemented would give TAXIR some relational powers, while retaining the fast search and efficient storage of the flat file system.

INTRODUCTION

At the University of Colorado in 1967 the authors invented the database management system called TAXIR (Estabrook and Brill, 1969; Brill, 1971), which today serves approximately a dozen user

Systematics Association Special Volume No. 26, "Databases in Systematics", edited by R. Allkin and F. A. Bisby, 1984, pp. 53–67. Academic Press, London and Orlando.

DATABASES IN SYSTEMATICS
ISBN 0 12 053040 6

communities in North America, South America and Europe, and is still undergoing development. TAXIR came to The University of Michigan in 1971 where it became interactive with dynamic storage allocation. In 1975 TAXIR was incorporated into the collection of programs supported by The University of Michigan for users of the MTS operating system. At Michigan TAXIR continued to grow in power and language, and, by 1977, descriptors had become completely modularized; report generation had become much more general; and aspects of the retrieval algorithm had been made faster (Brill, 1983).

In 1978 the MTS version of TAXIR was converted at the University of California at Berkeley to run under CMS, and in 1981 this version was taken to the Istituto per le Applicazioni del Calcolo in Rome, Italy where it was updated by the authors to the then MTS standard. In 1977, TAXIR went from Michigan to Brasilia, Brazil and thence to Rio de Janeiro where it has been used with the Brazilian flora inventory (Teixeira and Spiguel, 1980).

Other lines diverged earlier from the TAXIR phylogenetic tree. In 1970 the then current version was converted to run under OS on an IBM 360/67 for the USDA in Pullman Washington (Hudson *et al.*, 1971), and thence to Bari, Italy where it has been used not only for agriculture, but also for a floral survey of Italy (Anzaldi and Passerini, 1979; Pignatti, 1981). At the University of Colorado TAXIR continued to undergo some development and also experienced several name changes: STIRS (Klaphake, 1973; Chenhall, 1975); ENVIR, marketed as a proprietary package; and EXIR (Borsuk *et al.*, 1976), which travelled to the University of Quebec at Montreal where it learned to speak French (Legendre, 1978). TAXIR also travelled under the name of EXIR to the University of Southampton, UK, where it has been further developed by Bisby and White.

In recent years a number of TAXIR applications have been published in a variety of fields. In addition to those cited above, those in museums management and systematics include Peebles and Galloway (1981), LaRue (1979) and Moerman (1977), among others.

A USER'S VIEW OF TAXIR

What follows is a necessarily condensed and very incomplete sketch of TAXIR's power and language for building and querying databanks. Consult Brill (1983) for complete user's instructions and description.

1. Creating a Databank

A databank is a set of items corresponding to the things in some collection of interest, such as the specimens of a museum, publications reporting chemicals in plants, type specimens, etc. Information describing the items is structured into descriptors. A descriptor partitions the set of items into mutually exclusive subsets called descriptor states, or just states. For example, if the items represent specimens, the descriptor MONTH OF COLLECTION will have the states JAN, FEB, etc., and will partition the specimens so that the state JAN will have as members all the specimens collected in January, and so on.

A user creates a databank by defining the relevant descriptors. An item is entered by listing for each descriptor the states to which the item belongs. In addition to the descriptors defined by the user, TAXIR assigns sequential integer numbers, beginning with one, to the items as they are entered. This additional descriptor is called ITEM #. A databank can be thought of as a two-dimensional matrix of items vs. descriptors, with exactly one descriptor-state name at each coordinate. Thus, a TAXIR databank is called a "flat file" because of its two dimensional nature (Fig. 1).

There are three types of descriptors in TAXIR: FROM-TO; ORDER; and NAME. FROM-TO is for numeric data. In addition to a descriptor name, the user supplies a range and an increment, e.g.

	Family	Genus	Species	Habitat
Plant 1	Fam 25	Gen 392	Sp 106	Hab 11
Plant 2	Fam 31	Gen 112	Sp 82	Hab 4
Plant 3	Fam 98	Gen 45	Sp 16	Hab 6
Plant 4	Fam 16	Gen 309	Sp 30	Hab 8

Fig. 1. Some of the flat descriptors in a phytochemical databank. The columns are labelled with the names of the flat descriptors. The rows are labelled with the item numbers of the plants in the databank. In each cell of the flat file is stored a coded value representing the state of descriptor x to which the item y has been assigned. This value is, in fact, a pointer to the name of that state in a dictionary. A boolean expression involving flat descriptors invokes a search of this structure, resulting in a list of the item numbers of the plants that satisfy the search request.

YEAR (FROM 1750 to 1990 BY 1). ORDER is for alphanumeric data where the small complete set of states is known in advance. In addition to a descriptor name, the user supplies a fixed ordered list of states, e.g. MONTH (ORDER, JAN, FEB, MAR, APR, MAY, JUN, JUL, AUG, SEP, OCT, NOV, DEC). NAME is for alphanumeric data where the set of states is not known in advance. The user supplies only the name, e.g. COLLECTORS (NAME). Any state name listed during data entry becomes valid until deleted and thus the list of states may grow or shrink throughout the life of a databank.

For each descriptor, the valid state names are held in a dictionary (Fig. 2), which is used to scan user queries and prepare print-out.

Fam 15	Rosaceae
Fam 16	Rubiaceae
Fam 17	Araceae

Chem 41	HCN glycoside
Chem 42	coumarin
Chem 43	theobromine

Fig. 2. Fragments of a family dictionary and a chemical dictionary. To the left of the names in the family dictionary are the dictionary addresses to which the family values in the flat file (Fig. 1) are pointing. Likewise, the names in the chemical dictionary are labelled with the dictionary addresses to which the entries in the Plant/Chemical Pointer Structure (Fig. 3) are pointing.

Descriptors with many state names in common can share common dictionaries. Thus, state names occur at most once in the TAXIR internal storage structures. In addition to the states defined by the user, TAXIR automatically assigns each descriptor a state called UNKNOWN for missing information.

2. *Retrieving Data*

A TAXIR query consists of two parts: a list of descriptors; and a boolean expression. The *simplest* boolean expression has the form, "desc relop ds", where: desc is a descriptor name; relop is one of the relational operators $=, \neq, >, \geqslant, <, \leqslant$; and ds is a descriptor-state name of descriptor, desc.

The boolean operators NOT, AND or OR may be used, with parentheses nested to any depth, to combine boolean expressions of the simplest form into more complicated ones. Some examples of Boolean expressions in TAXIR are:

GENUS = Planta
YEAR < 1936
NOT COLLECTOR = SMITH and (MONTH < MAR or MONTH > SEP)

In the TAXIR language, a query has the form:

QUERY Descriptor List * Boolean Expression *

The information is printed one item at a time, for each item that satisifies the boolean expression. The printed information consists of the names of the states (of the descriptors on the query descriptor list) to which that item belongs. The order in which these groups of state names are printed is determined by the states of the first descriptor on the query descriptor list; within ties, by the states of the second descriptor on the query descriptor list, etc. Items in the same states for all descriptors on the query descriptor list are counted and when the count exceeds one, the count is printed in addition to the descriptor state names. The states of a NAME descriptor are ordered alphabetically; those of a FROM-TO descriptor are ordered numerically, and those of an ORDER descriptor are ordered in the order specified when the descriptor was defined. The descriptor list can be enriched by a number of formatting symbols to specify a wide variety of report formats, including indented taxonomic hierarchies, formal nomenclatural citations, address labels, invoices, etc.

For the special case of databases that can be reasonably construed as flat files in a TAXIR databank, TAXIR is very fast to discover the subset of desired items, and becomes relatively faster as the complexity of the boolean expression increases (Kahn, 1975). TAXIR's efficient algorithms, described by Brill and Estabrook (submitted), depend very heavily on the flat-file structure of its databanks. Thus, in spite of its efficiency, ease of use, and general reporting capabilities, one shortcoming of TAXIR is that often databases are not flat.

FLATTENING A FILE

It is often possible to flatten an "almost flat" data structure to make a TAXIR databank. There are basically four methods for doing this. As in Estabrook (1979), consider a nomenclatural databank of plant type specimens that is completely flat except for the author descriptor. Most of the basionyms have a single author, but some basionyms have two, three, or rarely, more than three authors.

1. Method 1

Define the descriptor, AUTHORS, to be of type NAME. If a basionym has two authors, A. B. Jones and J. R. Smith, then the state of the AUTHORS descriptor for this basionym can be "A. B. Jones et J. R. Smith". This is perfectly legal TAXIR and it flattens the file.

But boolean expressions involving AUTHORS become a little awkward. To find all the basionyms authored by J. R. Smith, it is not enough to use the boolean expression: "AUTHORS = J. R. Smith". This would find only basionyms for which J. R. Smith is the sole author. Instead, we give the command: "SHOW AUTHORS CONTAINS Smith". This generates a list of vocabulary entries that include the name Smith. It may look something like this:

6. AUTHORS CONTAINS Smith
214. A. B. Jones et J. R. Smith
98. E. N. Smith
45. J. R. Smith
150. J. R. Smith et Q. A. Perkins
101. L. O. Smith

Among the other Smiths, J. R. appears three times. Now we can compose a boolean expression to find all his basionyms:

AUTHORS = A. B. Jones et J. R. Smith or J. R. Smith
or J. R. Smith et Q. A. Perkins

A further difficulty comes when we generate an author index to the whole type collection. The reply to the command: "QUERY (AUTHORS, TITLE, ETC.) * ALL * " is ordered alphabetically on AUTHORS and most of the J. R. Smith basionyms are among the J. R.s, but not the basionyms by A. B. Jones et J. R. Smith. They will be found among the A. B.s.

In the present example, where basionyms have only one or a few authors, or where authors rarely collaborate, or where the order of the authors for a single basionym is important, the penalties are not severe. It is reasonable to use TAXIR for such an application.

An example for which Method 1 should not be used might be with a study of a natural habitat, say a meadow, that has been divided into square plots, which correspond to the items of a databank. Each plot may contain an average of 50 plant species, and each species may live in quite varied company from plot to plot. In this databank, plots are to species as basionyms are to authors, except that, if we flattened the file with method 1, the states of the descriptor PLANT NAMES would typically contain more different names than the states of the descriptor, AUTHORS, and would very likely each exceed TAXIR's limit of 112 characters per state, even if we abbreviate or code these names. Also, because of the number of distinct combinations of plants on plots, the size of the PLANT NAMES dictionary is potentially huge. To find all the plots with Planta alba, we would follow the same procedure as before, writing a boolean expression composed of perhaps hundreds of operands connected by ORs. This is not only tedious, but such an expression would not execute efficiently. Furthermore, a catalogue of plots ordered on PLANT NAME would be virtually useless because any one species, such as Planta alba, would be scattered throughout the print-out.

2. Method 2

Define two NAME descriptors, AUTHOR and COAUTHORS, declared to be equivalent, so that they share a single dictionary. For a basionym with a single author, we enter the author's name in AUTHOR and enter UNKNOWN in COAUTHORS. For a basionym with multiple authors, we enter as many items as there are authors of the basionym. These items are duplicate in all descriptors, except for each item, we enter a different author's name in AUTHOR and the remaining authors in COAUTHORS.

A basionym by A. B. Jones and J. R. Smith is represented by two items. In one item A. B. Jones is AUTHOR and J. R. Smith is COAUTHORS. In the other item the roles are reversed. An author index is produced by the command:

QUERY (AUTHOR, COAUTHORS, TITLE, ETC.) * ALL *

Each basionym appears once under each of its authors, accompanied by coauthors when they occur and by a blank space when there are none.

But the concept of an item has been violated. There is no longer a one-to-one correspondence between items in the data bank and the entities they represent. We must pay for the entry and storage of redundant information. The item counts supplied by TAXIR may not be correct. For example,

HOW MANY WITH AUTHOR = J. R. Smith or A. B. Jones *

will count too many; while

HOW MANY WITH AUTHOR = J. R. Smith *

will count correctly. Mildly strange behaviour may result from this basic violation of the item concept. E.g.,

QUERY (GENUS, SPECIES, AUTHOR, COAUTHORS) *
GENUS = Planta and SPECIES = alba *

might give this reply:

Planta alba A. B. Jones J. R. Smith
Planta alba J. R. Smith A. B. Jones

These occasional anomalies are worth it if your principal intention is to produce an author index. Moreover, boolean expressions are easier than in method 1: the SHOW statement is not needed, and AUTHOR = J. R. Smith correctly specifies all the J. R. Smith basionyms with a correct count. But if your principal intention is to correctly cite basionyms in their nomenclatural entirety, Method 1 is preferred. Finally, the data structure of plots and plants is no better flattened by Method 2 than by Method 1.

3. Method 3

Define three descriptors: AUTHOR 1, AUTHOR 2, and AUTHOR 3. We make them equivalent NAME descriptors so they will share a single dictionary. For the A. B. Jones et J. R. Smith basionym, AUTHOR 1 is A. B. Jones, AUTHOR 2 is J. R. Smith, and AUTHOR 3 is UNKNOWN (to mean that there is no third author). To find all the J. R. Smith basionyms use the boolean expression: "AUTHOR 1 = J. R. Smith or AUTHOR 2 = J. R. Smith or AUTHOR 3 = J. R. Smith".

This method works to give accurate responses and correct counts. But you must set a limit, N, to the number of author descriptors. Are three author descriptors enough? You could ignore any authors beyond the first three, or lump them with AUTHOR 3 (as in Method 1).

If you set N larger, then boolean expressions involving authors become more complex. And we still have the same problem with the author index that we had in method 1; many author names do not appear together in alphabetical order. Three author indexes, one based on each of the three author descriptors would be workable but cumbersome. Unless author order is an important concept to preserve in the databank, which it may well be, Method 3 tends to be inferior to the other two methods, and gets rapidly worse as N increases. Plots and plants is not flattened graciously by Method 3 either.

4. Method 4

Consider a new example in which the items are plant specimens. Any specimen may be supplemented by photographs and these may be of the following limited kinds: black and white negative, black and white print, colour negative, colour print, colour slide, or X-ray. If a specimen were supplemented with only one of these kinds of photos, we could define an ORDER descriptor: PHOTO (ORDER, B&W NEGATIVE, B&W PRINT, COLOUR NEGATIVE, COLOUR PRINT, COLOUR SLIDE, X-RAY). But a specimen may be associated with any combination of kinds of photos and this prevents this structure from being flat.

One solution is to use Method 1. But we can take advantage of the fact that the states are fixed and limited in number. Instead of one descriptor for all kinds of photos, we define for each kind of photo a separate descriptor with states: yes, no. TAXIR can list information about specimens in tabular form, showing yes's and no's in columns headed by the names of kinds of photo:

NAME COLLECTOR #	B&W NEG	B&W PRINT	COLOUR NEG	COLOUR PRINT	COLOUR SLIDE	X-RAY
Planta alba						
107324	yes	yes	no	no	no	yes
107328	no	no	no	no	yes	no
107345	yes	yes	yes	yes	no	no

For some purposes, Method 4 would work to flatten plots and plants. Define each plant descriptor to have only one state (other than UNKNOWN), namely the name of the plant to which it corresponds. Now one can accurately and easily choose plots that satisfy logical conditions relating to the plants that occur on them and print out other (perhaps ecological, geographic, etc.) information about them.

DESCRIPTOR GROUPS

Consider a databank in which the items are plant species, and the information concerns the presence of chemicals of interest in each species as reported in publications. We might have two descriptors that take multiple values, CHEMICAL and SYNONYM. If a synonym index, but not a chemical index, is wanted then Method 2 for SYNONYM and Method 1 for CHEMICAL might be a good mix. If one wants both a synonym index *and* a chemical index, one could use Method 2 for both descriptors, but run the danger of over-inflating the databank with redundant data. The number of items needed for each species is the number of synonyms it has times the number of chemicals it contains. This requires redundant storage and it intensifies the side effects of violating the item concept.

In general, any number of descriptors can be flattened in the same databank, using the best method for each, as long as the descriptors involved are independent of each other. SYNONYM and CHEMICAL are independent in the sense that the choice of values for CHEMICAL in no way logically constrains the choice of values for SYNONYM, and vice versa.

However, suppose we also want to store in the databank information about the publications that report chemicals in plants, using descriptors, TITLE, DATE, PLACE and PAGE NUMBER. For a species with more than one publication, we must enter multiple values for the descriptors TITLE, DATE, PLACE and PAGE NUMBER. We must ensure that the parts of each citation remain in association, so that we end up with a set of citations, not just a set of titles, a set of dates, a set of places and a set of page numbers. We might call such an association of dependent descriptors a group.

Method 1 fails to preserve the inner association of the publication group, but Method 2 can be used. For each species instead of just one item, an item for each publication that cites it is needed. This works, but incorrect species counts can arise. Method 3 can be used provided the number of publications allowed for one species remains less than a small N. Method 4 is not applicable to this example, but it can be used to flatten some descriptors in groups where the number of states is small.

The publication group, because we allow it multiple values, is a bump on an otherwise flat file of species. But if we look at the publication group in isolation, it is a mini-databank, in itself quite flat. In each sub-item, there is but one title, one date, one place and one page number. What then if we give the publication group a bump of its own? Let us add the descriptor AUTHOR, permitting it multiple values in each publication.

TAXIR can still handle this. One way is to use Method 2 to handle the publication group as a whole, and Method 1 to handle the author descriptor. For each species, we enter one item for each publication (Method 2), and put the authors into the author descriptor (Method 1). We must pay Method 2 penalties for publications and Method 1 penalties for authors. We could also use Method 2 with Method 3, or Method 3 with Method 1, with varying degrees of success.

Bumps can be added to bumps indefinitely, and the above flattening methods can be applied to flatten any data file, however bumpy, and manage it with TAXIR. But TAXIR was designed to accept flat files. Applications that are naturally flat are the most appropriate for TAXIR. A little flattening is often acceptable, but the more flattening you do the more awkward and less efficient TAXIR becomes.

MULTIPLE DESCRIPTOR

Clearly, most desirable from the user's point of view would be for TAXIR to have the power to accept and use a descriptor with multiple states per item. A simple boolean expression with such a descriptor could be used to build more complicated boolean expressions, and the states of a multiple descriptor could be printed as part of a TAXIR query response. Although at present TAXIR cannot do this, we sketch below a preliminary design for it.

Consider a databank whose items are plant species described by ordinary flat TAXIR descriptors, such as FAMILY, GENUS, SPECIES, HABITAT etc. plus a multiple descriptor to indicate what chemicals of interest are known to occur in each species. We would like to ask:

QUERY GENUS, SPECIES, HABITAT * CHEMICAL=coumarin and FAMILY=Rubiaceae *

and get an alphabetized list of species in the Rubiaceae known to contain coumarin. Or we would like to ask:

QUERY GENUS, SPECIES, CHEMICAL * HABITAT=wet tropics and FAMILY=Rubiaceae *

and get an alphabetized list of species in the Rubiaceae from the wet tropics each associated with the chemicals of interest that it contains.

One way to implement this capability partially in TAXIR would be through the use of a Plant/Chemical Pointer Structure (Fig. 3). This consists of two things. The first thing is a list in plant item number order. The entries in this list are of two kinds: in the event that a plant contains only one chemical of interest, the entry is the address of the name of this chemical in a dictionary of chemical names; but if a plant is known to contain more than one chemical of interest, then the entry is the starting address of a linked list (the second thing) containing all the addresses in the chemical names dictionary of the names of all the chemicals in that plant.

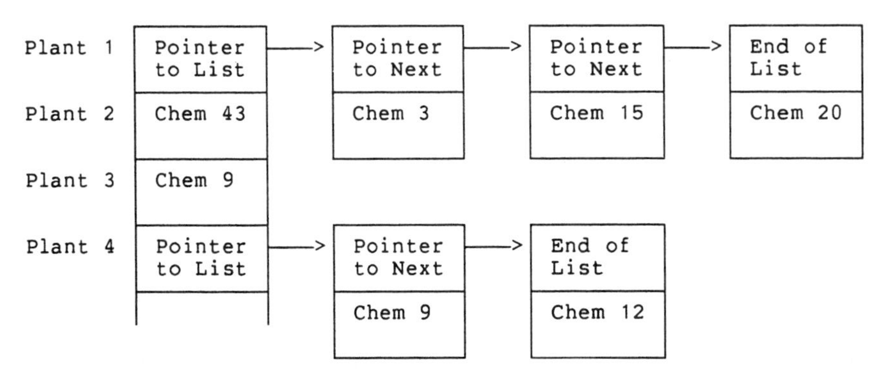

Fig. 3. Plant/Chemical Pointer Structure. Item number "Plant y" not only points to the row of the flat file (Fig. 1) containing the dictionary pointers for the flat descriptors, but also points in this structure (Fig. 3) to a list of dictionary pointers for the chemicals contained in plant y.

This structure would enable TAXIR to answer queries of the second type illustrated above. Standard TAXIR powers provide a list of plant species item numbers for the Rubiaceae from wet tropics, and the Plant/Chemical Pointer Structure provides the chemical name for each of the chemicals that occurs in each of these plants.

But the structure would be inadequate to answer queries of the first type illustrated above. To complete this task we need a Chemical/Plant Pointer Structure as well (Fig. 4). This is analogous to the Plant/Chemical Pointer Structure: the list is indexed like the chemical names dictionary; an entry is a plant item number in the event that that chemical occurs in only that plant; but in the event that many plants have this chemical, the entry is the address of the beginning of a linked list of plant item numbers.

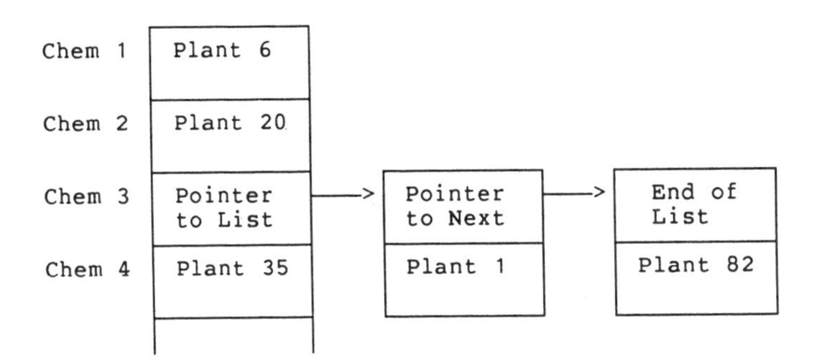

Fig. 4. Chemical/Plant Pointer Structure. The pointer "Chem x", which may appear one or many times as an entry in the Plant/Chemical Pointer Structure (Fig. 3), not only points to chemical x's name in the chemical dictionary (Fig. 2), but also points in this structure (Fig. 4) to a list of item numbers for the plants containing chemical x.

Now, TAXIR finds coumarin in the chemical names dictionary, enters the Chemical/Plant Pointer Structure with the dictionary address of this chemical, finds the plant item numbers of plants containing coumarin, and uses these in its retrieval algorithms to find coumarin-containing members of the Rubiaceae whose names and habitats it will print.

PREJOIN

Often, databases are neither flat nor graciously flattenable. Sometimes it is desirable to efficiently store repeating detailed information, such as the publication group in the databank of plant species containing chemicals, discussed above. In this case, publications could play the role of items in some other databank. In a databank whose items are species, the publication group could now be replaced by a single descriptor whose states are item numbers in the publications databank. In this databank, whose items are publications, we have descriptors such as TITLE, AUTHORS, PLACE, DATE, PAGE

NUMBERS, and PLANTS, of which AUTHORS and PLANTS would be multiple.

A multiple descriptor with Plant/Publication and Publication/Plant Pointer Structures would be needed for both the plant databank and for the publications databank. Of course, TAXIR would only need one copy of each. In addition, now that dictionaries exist for all the descriptors in the publications bank, the Plant/Publication Pointer Structure could contain publication item numbers instead of dictionary addresses. Queries could go both ways: searching in publications, but printing information about plants – "Plants from what habitats were surveyed for chemicals before 1940?"; and searching in plants, but printing information from publications – "Who has published information about plants containing coumarin from the wet tropics?"

Such interaction between related databanks is known in relational terms as joining the databanks. This interaction is possible when each databank contains a "key" to the other. In our example, these keys are the relation between plants and publications: a publication references a set of plants, and a plant is referenced by a set of publications. The relations "references" and "is referenced by" are inverses of each other. By storing the information represented by these relations, we can pass back and forth from one databank to the other. The general concept of relational database management is reviewed by Chamberlin (1976).

We call the proposed join capability for TAXIR a prejoin because, unlike the completely general relational model for database management, TAXIR's special case of "join" has to be prespecified at the time the databank is to be built, and special internal structures (multiple descriptor) must be built to anticipate its use. The general join operation in relational database management systems can be very expensive in time and space to execute. Although TAXIR would require that the user indicate, in advance of entering data into the databanks, his intent to join them in specified ways, this ensures a deliberateness and efficiency that is consonant with the use philosophy of TAXIR-built databases.

The implementational challenges presented by multiple descriptor and its corollary, prejoin, are great. For the sake of presenting the basic ideas here, these challenges have been ignored or minimized. However, the power that such implementation would provide to the world community of TAXIR users would be great. The authors hope that the personnel and financial resources to effect this implementation will be forthcoming in the near future.

REFERENCES

Anzaldi, C. and Mirri-Passerini, L. (1979). "Un esperimento di strutturazione di dati floristici e vegetazionali". Istituto per le Applicazioni del Calcolo [Pubblicazioni Ser. 3, no. 195], Rome.

Borsuk, R., Dyer, W., Hanley, J. R., Legendre, P., Rogers, D. J., Sim, B. A., Vincent, L. and Watt, D. (1976). "EXIR User's Manual". Information Sciences/ Genetic Resources Program, University of Colorado, Boulder.

Brill, R. C. (1971). "The Taxir Primer". Institute of Arctic and Alpine Research [Occasional Paper no. 1], University of Colorado, Boulder.

Brill, R. C. (1983). "The Taxir Primer, MTS Version". 4th edition. University of Michigan Computing Center, Ann Arbor.

Brill, R. C. and Estabrook, G. F. (submitted). Data subset selection by Boolean calculation. *Mathematical Biosciences.*

Chamberlin, D. D. (1976). Relational data-base management systems. *Computing Surveys* 8, 43–66.

Chenhall, R. G. (1975). "Museum Cataloging in the Computer Age", American Association for State and Local History, Nashville.

Estabrook, G. F. (1979). A Taxir Data Bank of Flowering Plant Types at the University of Michigan Herbarium. *Taxon* 28, 197–204.

Estabrook, G. F. and Brill, R. C. (1969). The theory of the Taxir accessioner. *Mathematical Biosciences* 5, 327–340.

Hudson, L. W., Dutton, R. D., Reynolds, M. M. and Walden, W. E. (1971). Taxir – A Biologically Oriented Information Retrieval System as an Aid to Plant Introduction. *Econ. Bot.* 25, 401–406.

Kahn, M. A. (1975). "A Benchmark of Three MTS Data Base Management Systems – MICRO, SPIRES and TAXIR" PRISM Associates, Ann Arbor.

Klaphake, E. W. (1973). "STIRS User's Manual". University of Colorado Computing Center, Boulder.

LaRue, J. I. (1979). Using a local computer for garden records management. *Bull. Am. Ass. bot. Gn Admin.* 13, 16–18.

Legendre, P. (1978). "EXIR 3.0, Guide de l'Usager". Universite du Quebec, Montreal.

Moerman, D. E. (1977). "American Medical Ethnobotany, A Reference Dictionary". Garland Publishing, New York.

Peebles, C. S. and Galloway, P. (1981). Notes from underground; Archaeological data management from excavation to curation. *Curator* 24, 225–250.

Pignatti, S. (1981). "Checklist of the Flora of Italy with Codified Plant Names for Computer Use". Consiglio Nazionale delle Ricerche, Rome.

Teixeira, A. R. and Spiguel, C. P. (1980). Banco de dados do Programa FLORA do CNPq, sobre plantas medicinais e farmacologia de produtos naturais. *Ciênc. Cult. S. Paulo* 32, 48–58.

6 A Concept for a Machine-readable Taxonomic Reference File

M. N. DADD

BIOSIS UK Ltd, 44 High St., Boston Spa, W. Yorks LS23 6EA, UK

and

M. C. KELLY

BIOSIS Inc., 2100 Arch St., Philadelphia, PA 19103, USA

Abstract: BIOSIS and *Zoological Record* have been working together for over 5 years developing the concept of the Taxonomic Reference File – a computerized collection of organism names and associated data. Its objective is to provide the scientific community with an on-line, interactive tool for sharing taxonomic information. The original concept was aimed at bringing consistency to the taxonomic vocabularies and classification schemes of abstracting and indexing services. More recently an expanded concept has evolved in response to feedback from demonstrations and discussions with taxonomists from many parts of the world. It has been apparent from the outset that community participation in the design and development of the system is essential. An Advisory Group is being established which will be invited to contribute its views at each stage of development. In addition, the scientific community can have an important role in building the files. We hope to be able to make use of respected sources of systematic data, especially those which already exist in computer-readable form. This will be supplemented by material from the current literature passing through the BIOSIS and *Zoological Record* systems. When the file is available, users will be able to search it on-line in a manner similar to that now used for searching bibliographic files using any organism name, authority, or other descriptive data or any combination of these. In addition, approved users will be able to record their scientific findings in special user comment fields associated with the file,

Systematics Association Special Volume No. 26, "Databases in Systematics", edited by R. Allkin and F. A. Bisby, pp. 69–78. Academic Press, London and Orlando.

DATABASES IN SYSTEMATICS
ISBN 0 12 053040 6

and these will be reviewed and amended periodically by cooperating authorities. It is also intended that users will be able to copy portions out of the main file and into a separate workspace area of the system in which they can add new information or make changes to suit their own needs.

HISTORY OF THE PROJECT

BIOSIS and *Zoological Record* have been working for a number of years to develop a concept for a computer-readable reference file in taxonomy.

BIOSIS is a private non-profit (the US equivalent of "charitable") organization, established in 1926 under the auspices of the US National Research Council, the American Association for the Advancement of Science and what was at that time the Union of American Biological Societies. It has operated continuously for some 57 years abstracting, indexing and processing the spectrum of the life sciences research literature in all of its ramifications. BIOSIS has expanded with the growth of the biological literature and has, we feel, developed a leading position in the area of information systems, and in the creation and searching of massive bibliographic files. A history of BIOSIS is given by Steere (1976).

The *Zoological Record* was started in 1864 and has been published continuously since 1886 by the Zoological Society of London. A history is given by Edwards (1976). As a result of a 1980 agreement with the Zoological Society of London, BIOSIS is now joint publisher of the *Zoological Record*. Since this agreement BIOSIS has been responsible for management and finance of the *Zoological Record*, the staff employed by the Society on the production of the *Record* having transferred to a new BIOSIS subsidiary, BIOSIS UK Ltd, while the Society has remained responsible for broad editorial matters. The *Record* continues to be produced on a cost-recovery basis as a service to the biological community.

For many years we have recognized a problem in "verifying" the names of organisms used in the literature that we, as secondary services, attempt to index. While the determination of authoritative nomenclature is outside the scope of our organizations, the need to apply consistency in the use of names remains. We use the term "verification", therefore, to indicate that we regard the name as recognizable to systematists and that we are able to assign it to the hierarchy we use in preparing our taxonomically-oriented indexes.

Many potential problems have been and are resolved by the use of cross-references between synonyms, alternative classification schemes, etc. However, while this solution, leaving the user to track down the location of information, works in the printed publication, it is less helpful in the on-line environment where the user, especially the non-taxonomist, may not even realize that the problem exists. In order to be able to consistently "verify" names, we have built up over the years very large card indexes which have become increasingly unwieldy. Coupled with the need to provide a better means of locating information through the name of an organism when using on-line search services such as Dialog, we recognized the need to find a more efficient system.

This history was common to BIOSIS, *Zoological Record* and a number of other secondary services dealing with biological literature, including *Chemical Abstracts*, *Bulletin Signaletique*, etc. The forum which brought us together and enabled discussion to take place was ICSU-AB – the International Council of Scientific Unions Abstracting Board which, despite its name, is not part of the ICSU family but affiliated to it. This organization consists of representatives of the various secondary services, the scientific unions and national members. It was through the mechanism of its "Life Sciences Working Group" that we first began discussion of our mutual problem of "verification".

As a result of these discussions, in 1975 BIOSIS and *Zoological Record* began a project to study the feasibility of what we then chose to call an International Taxonomic Reference File, or TRF. The stated purpose was:

> to investigate the feasibility of producing a reference tool which would make it possible for information services to verify, for indexing and retrieval purposes, the names of biological organisms referred to in scientific literature.

In addition, the tool was to indicate the position of the taxon in an hierarchical classification and to "translate" between different classifications where more than one system was involved. Although this project was largely supported by our own funds, we did receive some assistance from the British Library Research and Development Department towards travelling expenses. As part of this feasibility study, and following many discussions between representatives of BIOSIS and the *Zoological Record*, we agreed on respective areas of expertise. BIOSIS took responsibility primarily for the design of a machine-readable system that would incorporate the necessary information, and *Zoological Record* took a major role in identifying the kinds of information that would be appropriate for such a file.

Following on from the study, work began on the development of a small test file, intended to include about 40 000 records. BIOSIS did the necessary design and programming work and files were provided by *Zoological Record* for demonstration purposes. This was demonstrated in seminars at BIOSIS in Philadelphia and at a workshop, again supported by the British Library, held in London in January 1979 and attended by representatives from organizations such as the British Library, British Museum of Natural History, The Royal Chemical Society, the Commonwealth Agricultural Bureaux, Excerpta Medica, Forschungs-Institute Senckenberg of West Germany, Institute of Terrestrial Ecology, and The Royal Botanic Gardens at Kew. The results of the feasibility study, and the workshop which followed it, were reported by Dadd (1976, 1979).

From these demonstrations and from the workshop came valuable suggestions for improving the system design and for making the file one that would be more acceptable to the biological community at large; one that might serve them not only indirectly through the secondary services, but also directly. Against the background of those comments and suggestions the system design for the TRF has been modified to provide the scientific community with an on-line, interactive tool for sharing taxonomic information.

SYSTEM OVERVIEW

The Taxonomic Reference File (TRF) will have four main components:

(1) The Taxonomic Data File will include organism names plus nomenclatural data. This file will be based on names. Each name which has appeared in the literature will have its own entry and will be assigned a unique, identifying number. Within this list of names, key names will be designated as preferred for the purposes of BIOSIS and *Zoological Record* indexing. All names will be linked with their alternate versions, such as synonyms, spelling variants, and misidentifications. An indication of the relationship of these names will be given and provision will be made for annotating disputed nomenclature. Where a major nomenclatural work, or works exist(s) for the

group, an indication will also be given of how that work treats the name.

(2) The second component of the TRF, the Hierarchy File, will consist of taxonomic classification schemes. It is planned that the Hierarchy Files will carry multiple schemes to accommodate differences in classification. Each name in the TRF index will carry information regarding its taxonomic level and will be linked to one or more of the taxonomic trees in the Hierarchy File. Since an important application of this Hierarchy File will be to provide consistency in indexing for secondary services, the BIOSIS and *Zoological Record* indexing schemes will be included as one (or two, where we differ), of these taxonomic trees. Other cooperating services will also be included.

(3) The third component of the TRF will be the Related Data Files. Multiple files are planned in order to accommodate different kinds of data. For example, descriptive data about each organism (such as common names, geographical and geological range, type specimen and locality) can be included here. The multiple-file approach provides an opportunity for future expansion of the TRF to meet the diverse needs of scientists in fields related to taxonomy. It is also possible to allow for supplementary data to be included where available (such as host/pest data, endangered species legislation, derivative chemicals, toxin antidotes, etc.). Different types of data may be included for different kingdoms of organisms, as appropriate. By providing for supplementary data elements of this nature, we hope that the usefulness of the file will be expanded to serve a broader audience and that sufficient flexibility will be provided in the system to accommodate potential new directions in taxonomic research.

(4) The fourth component of the TRF will be the Bibliographic Data File. This file will serve as a link between entries in the TRF and the related bibliographic information to be found in BIOSIS's existing printed and computer-readable products, *Biological Abstracts*, *Biological Abstracts/RRM* (and its predecessor, *BioResearch Index*), *BIOSIS Previews*, *Biological Abstracts on Tape*, and the *Zoological Record*.

Information in the four component files will be linked together in the computer by the unique, identifying number which is assigned to each organism name. In this way, it will be possible for users to pull together data from each of the files for a given organism name.

In addition to this computerized collection of organism names and associated data, and critical to the TRF's success, is the interactive computer system which will allow users to look at the data, add to or comment on them when authorized, create individually tailored subsets of the data for local use, and share findings with other scientists, all through an on-line computer network.

At the simplest level, users will be able to search the file in a manner analogous to that now used for searching bibliographic files. This means that users will be able to interrogate the file using any organism name, authority or other descriptive data or any combination of these. In addition, certain approved users will be able to record their scientific findings into special user-comment fields associated with the file. Each comment entered in this fashion will carry a clear identification of its author and will be reviewed and amended periodically by cooperating authorities. To protect the TRF from unauthorized changes, users will *not* be able to change directly information already on the file.

In order to make the TRF more useful to taxonomists, the user will also be able to copy portions of the TRF out of the main file and into a separate workspace area of the system. Here the user can add new information or make changes to suit his needs. This facility would enable the user to create a modified version of the TRF without affecting the contents of the main file. This modified version might be useful for example for producing museum catalogue cards, or the user might copy off a subset to put on a local mini- or microcomputer, for further manipulation.

An additional feature proposed for the TRF system is a limited message exchange facility. Here users could record messages alerting colleagues to new information in their areas of interest. To do this a user would enter a message and indicate who should receive it. When the designated recipient signed on to the system, the message would be displayed for him and then erased.

Other features which are planned for the system include batch output in such forms as computer listings and tapes. This technique could be used to produce subset listings on selected data elements or organism groups, including information which was sorted to meet special needs. Although no current plans exist for producing printed publications from the TRF, this represents a potential future output, and a mechanism for producing photocomposed output is included in the TRF system design.

BIOSIS AND THE COMMUNITY'S ROLE

Given the scope and sheer volume of taxonomic data, it is impractical to attempt a project like the TRF without the cooperation of the scientific community. BIOSIS will provide a focal point for the coordination of information via this system and the system design itself. It is BIOSIS's wish to involve the community directly in the building and maintenance of the TRF database.

We anticipate that there will be several levels of community involvement with the TRF. To begin with, an advisory group of knowledgeable individuals from around the world is being put together. This group will make recommendations regarding current TRF plans, will evaluate operational feedback as we proceed with implementation, and will give advice on future directions for the TRF.

The scientific community will, we hope, also serve as a source of data for the file. BIOSIS plans to make use of respected sources of systematic data, especially those which already exist in computer-readable form. To accomplish this, individually tailored programs will be written to take existing data files in nomenclature, taxonomy, or related fields and reformat them to meet the specifications of the TRF. We already have some experience with this technique. A demonstration TRF file of some 40 000 entries was created by batch processing a computer-readable file of taxonomic data from *Zoological Record*. In addition, the technique has been used to incorporate into a Pilot File, to be discussed later, retrospective bibliographic data which has appeared in the BIOSIS printed and computer-readable products. We hope to cxpand on this approach to include other files existing in the community.

Once the base file for the TRF has been built, we would again look to the community for assistance in maintaining its contents. BIOSIS hopes to obtain cooperation from authorities in reviewing and maintaining information in assigned sections of the file. For example, a cooperating authority might be responsible for reviewing and maintaining information on the geographical and geological ranges of a specified group of organisms. BIOSIS itself can contribute to maintaining the currency of the file, by making use of the information to be found in its bibliographic databases. We plan to write computer programs which will automatically identify and input relevant systematics data as they enter the BIOSIS and *Zoological Record* publication databases. When a new taxon is identified in the publication database, it will result in an entry being added to the TRF.

In addition to using cooperating authorities and the BIOSIS database, we also plan to look to the users of the TRF as a source of new taxonomic information. In order to make the TRF a dynamic reference tool, we propose that certain approved users be permitted to enter their scientific findings directly into the TRF in special user-comment fields. This would be analogous to the "Letters to the Editor" column in a journal. Each comment entered in this manner will carry a statement of authorship and will be periodically reviewed, and possibly amended, by cooperating authorities. Of course this capability would not be available to all users – the general user would only be allowed to search the file and examine its contents.

IMPLEMENTATION

The Taxonomic Reference File (TRF) is a mammoth undertaking, one which, because of its dynamic nature, will never be completed. BIOSIS is well aware that others have attempted activities similar to the TRF, some of which, for various reasons, have never been completed. However, these earlier attempts, in addition to highlighting the difficulties, also point out strong interest in a computerized reference file in taxonomy.

BIOSIS recognizes that for the TRF to succeed, care must be taken in moving beyond the planning or systems design stage and into the operational mode. We have begun to recruit the advisory group. This is a critical first step; we feel that it is essential to get input from experts who can make recommendations regarding details of the current TRF plans, who can evaluate operational feedback as we proceed with implementation and who can suggest future directions. Also, we are in the process of exploring possible sources of funding. There are many aspects of this project which BIOSIS is able to support; however, outside funding will be essential if the TRF is to fulfil its potential within a reasonable time. Finally, we have prepared specifications for and have begun work on a TRF Pilot Project which will give us an opportunity to conduct a real, hands-on test of the full TRF concept.

As a first step in the Pilot Project, we are in the process of building a start-up database which will contain some of the taxonomic data elements planned for the full TRF. At the outset we will include organism names, an indication of their hierarchical relationships, and limited taxonomic data. Additional data will be available in the form of bibliographic citations and abstract texts from BIOSIS publications. This start-up database will serve as a skeleton file for use in implementing the pilot system.

Practical realities have necessitated our selecting a subset of all living organisms to use in building this start-up file. We have chosen bacteria as encompassing a manageable number of taxa. There are approximately 2300 bacteria, with some 37 000 names including synonyms, and some 36 000 taxonomic papers cited in our retrospective files from 1926 to date. Bacteria seem an appropriate choice not only because of their manageable numbers, but also because they have implications for other taxonomic groups, for example, in their role as pathogens of both plants and animals.

We expect that this start-up file will be completed shortly. Keyboarding is already complete, and proofing and correction are underway for the start-up file components. The bibliographic start-up file contains approximately 8000 citations and abstracts pertaining to bacterial taxonomy from 1969 to date. The skeleton for the taxonomic data file has been built using a BIOSIS authority file containing 8000 bacterial names with associated taxonomic data. Four different hierarchy files are also under development. Two derive from BIOSIS classification/indexing schemes, and two are from respected community reference works. Provision is being made to include a related data file. We will be working with the scientific community regarding the cooperative use of existing collections of related bacterial data.

In addition to these file-building activities, work is also underway to assemble a small user group to take part in an on-line, interactive test of the pilot system. Unlike the advisory group which will include representatives from botany and zoology as well as microbiology, this user group will be limited to 15–20 scientists with expertise in bacteriology. The user group will play an important role in testing the utility of the TRF contents and its format as well as the on-line system features. BIOSIS will be acquiring the telecommunications equipment necessary to run this pilot system test.

From this Pilot Project there is much to be learned about how users will interact with a taxonomic data file. Our experience with bibliographic reference files is useful but not sufficient to predict the best ways in which data can be organized and user—system interactions can be structured. The Pilot Project is meant to test the full TRF concept and will include most of the system features planned for the TRF. While it is true that the database for the Pilot Project will be limited to bacteria, the system test is intended to help us determine a workable data structure for the larger TRF database. The test will also allow us to assess user needs and the nature of future community participation. The Pilot Project is designed as an educational process, one which will allow us to begin work on the full TRF with greater confidence in its feasibility and its community acceptability.

REFERENCES

Dadd, M. N. (1976). "Standardisation of biological names. An investigation of methods for use by information services". British Library Research and Development Department Report no. 5347.

Dadd, M. N. (1979). "A machine-readable taxonomic reference file. International Workshop January 29th—31st, 1979". British Library Research and Development Department Report no. 5548.

Edwards, M. A. (1976). The Library and scientific publications of The Zoological Society of London: Part II. *Symp. Zool. Soc. Lond.* 40, 253—267.

Steere, W. C. (1976). "Biological Abstracts/BIOSIS. The first fifty years". Plenum Press, New York and London.

7 The European Taxonomic, Floristic and Biosystematic Documentation System — An Introduction

V. H. HEYWOOD, D. M. MOORE, L. N. DERRICK, K. A. MITCHELL and J. van SCHEEPEN

Department of Botany, University of Reading, Reading RG6 2AS, UK

Abstract: The European Science Foundation's European Taxonomic Documentation System is financed by ten member countries' Research Councils. Its aims are to provide a floristic, biosystematic and taxonomic information system for the vascular plants of Europe. The foundation of the database is the systematic geographical, ecological and chromosome information cited in *Flora Europaea*. The processing of this information into searchable computer files represents the first and current phase of this project. Updating of information in these fields and the creation of additional fields on such categories as illustrations, conservation status, phytochemical information, economic importance and phytosociology will be covered in the second phase of the project.

It is envisaged that enquirers will be able to search the database by using a Viewdata terminal. In this way the user will receive information on the status of a taxon by its colour on the screen. For example, accepted names will appear in white, synonyms in green and post-*Flora Europaea* names will be in yellow. It is planned to make simple on-line searches from the terminals, and more complex searches will be possible with the aid of an electronic message facility. It is planned to produce an updated synonymic catalogue of the European flora as one of the first products. Biosystematic surveys of particular groups and various checklists are also planned. There will be draft surveys produced to test consumer reaction. It is not intended to revise the data as they accumulate, but taxonomic revisions of particular groups will be commissioned when and as funds are made available and the original *Flora Europaea* database will be modified accordingly. In the meanwhile, up-to-date information will be provided for the user to process further as it is needed.

Systematics Association Special Volume No. 26, "Databases in Systematics", edited by R. Allkin and F. A. Bisby, 1984, pp. 79–89. Academic Press, London and Orlando.

DATABASES IN SYSTEMATICS
ISBN 0 12 053040 6

INTRODUCTION

The proliferation of taxonomic and floristic data on European plants in the post-World War II period led to the initiation of the *Flora Europaea* project in 1955. The history of this project has been reviewed by Heywood (1957, 1958, 1978) and Webb (1978). From the point of view of floristics and plant taxonomy, the five published volumes of *Flora Europaea* (Tutin *et al.*, 1964–80) constituted an overall synthesis on a continental scale of the flowering plants and ferns known to occur as native or subspontaneously in Europe, and provided European biologists and other users with a database for future research. In a wider sense, however, *Flora Europaea* laid the basis for European cooperation in plant taxonomy for decades to come and also served as a model for collaborative efforts in other fields of endeavour.

In no sense was *Flora Europaea* intended to be a definitive work, although it could be accepted for many purposes as a standard treatment. Moreover, the editors were fully conscious of the fact that one of the effects of its publication would be to stimulate further research on the taxonomy, nomenclature and distribution of European plants. Although they did not envisage themselves preparing new editions or supplementary volumes of additions or corrections, apart from minor changes when individual volumes were reprinted, a pilot scheme for accumulating and editing additions and corrections, using the family Caryophyllaceae, particularly the genus *Silene*, as an example, was initiated in 1979 at Cambridge under the direction of S. M. Walters with support from the *Flora Europaea* Trust Fund of the Linnean Society of London.

The volume of data published since the appropriate volumes of *Flora Europaea* has been such that it soon became evident that there was a serious danger of European taxonomy and floristics reverting to the position it found itself in when the Flora was initiated.

In 1975 both the Committee of European Science Research Councils (ESRC) and The Royal Society of London proposed the setting up of *ad hoc* groups to consider European projects in biological taxonomy and systematics and long-term coordination of efforts and policies of the Research Councils on such matters. They were referred to the European Science Foundation (ESF) which recommended the establishment of an *ad hoc* group on Biological Recording, Systematics and Taxonomy by the ESRC, which by then had become a Standing Committee of the ESF, in close cooperation with The Royal Society of London. The *ad hoc* group was divided into two sections – one for animal taxonomy and one for plant taxonomy – and met six times between 1976 and 1980 when it was formally disbanded. Its first task was to examine the problems of national resources and needs (both human and physical), multinational projects, aid programmes and taxonomic services (ESF, 1977). In addition the Botany Section considered the current state of knowledge and research on the taxonomy of European plants and identified a number of areas as deserving urgent and immediate attention. One of these was a "European Floristic Information System" and another related project was the "Coordination of research and information on the biosystematics of European plants". The Floristic System was planned to encode and store in computer-retrievable form specified fields of information on European plants derived from *Flora Europaea* which would act as a base-line. The data files would subsequently be updated in the light of additional information published since the preparation of *Flora Europaea* and additional files would be created for fields of information not covered by *Flora Europaea*.

The Biosystematics Programme was intended as a means of coordinating biosystematic research on a European basis. It was noted that, unlike the situation with floristic data, which are coordinated and synthesized in the form of Floras, there is no tradition or mechanism for the gathering, coordination and synthesis of biosystematic data, nor a means of publishing synopses of the processed data in a systematized manner comparable with Floras, although various chromosome indexes are available.

The problem is not new, nor is it one that refers only to Europe. Attention was focused on it at the International Botanical Congress in Montreal in 1959 and was behind the establishment of the International Organization of Biosystematists (IOB, later International Organization of Plant Biosystematists – IOPB) in 1960. One of the aims of the IOB (and later IOPB) was to establish a registration centre for the assembly and dissemination of biosystematic data (Heywood, 1962). Apart from the publication of "Biosystematic Literature" (Solbrig and Gadella, 1970), which was a compilation of literature for the period 1945–64, little progress has been made.

Subsequently it was decided to unite the European Floristic System and the Biosystematics Programme as a single project which was submitted to the European Science Foundation as Project I of an Additional Activity in Taxonomy under the title "European Floristic, Taxonomic and Biosystematic Documentation System". This was approved by the General Assembly in November 1979 and it was decided to base the project at the University of Reading in the Department of Botany under the direction of Professor V. H. Heywood, if suitable arrangements could be made with the University. A contract was signed between the European Science Foundation and the Unversity of Reading in March 1981 although the starting date was postponed until November 1981 for technical reasons. A Steering Committee for the project was nominated by the ESF, comprising of representatives of the countries involved in financing the project (currently Austria, Belgium, Denmark, France, Germany, Ireland, Italy, Norway, Sweden, Switzerland, United Kingdom). The role of the Steering Committee, which meets at least once a year, is to review the budget proposals for the following year, scrutinize the qualifications of candidates for appointment to posts financed by the Project, and monitor progress. The Steering Committee met in Ghent in June 1980, and discussed the implementation of the programme in detail. It was decided that the geographical coverage of the "Documentation System" should be compatible with the Mediterranean Check-list. Professor D. M. Moore was nominated as Associate Project Leader. A second meeting of the Steering Committee was held in May 1982 in Reading.

THE DOCUMENTATION SYSTEM

The aims of the project were originally defined as follows:

(1) To devise and establish a European information system for the accumulation, collation, revision and selective publication of new data referring to the taxonomy, floristics, systematics and biosystematics of European Flowering Plants.

(2) To provide a databank, to be continually updated, on all the vascular plants reported from Europe.

(3) To provide geographical information (within and outside Europe) on all such taxa recorded from Europe.

(4) To provide ecological and phytosociological data on all such recorded entities.

(5) To list literature references pertinent to all such taxa.

(6) To indicate the conservation status of all such taxa.

(7) To provide a databank of biosystematic information on all such taxa (also at infraspecific level), e.g. chromosome numbers, origins of material, karyotype details, pairing behaviour at meiosis, hybrids, genecological information, chemical information, etc.
(8) To prepare from these data biosystematic surveys or monographs of selected groups of high scientific and/or economic interest.
(9) To devise and experiment with novel methods of presenting such information.
(10) To highlight deficiencies in our knowledge of particular groups with a view to suggesting future areas of research.
(11) To provide and publish an up-to-date annotated check-list of the European flora and other selections of the information stored in the system as appropriate and according to demands.

The project was planned to have two distinct phases. The first was to consist of the construction of a suitable file structure for the basic data in *Flora Europaea* (names, authorities, places of publication, geographical distributions, etc.), the inputting of these data into the computer and preparatory work in selected data fields not covered by *Flora Europaea*.

Phase II was to consist of the updating of the information in the fields covered by *Flora Europaea* from current and post-*Flora Europaea* literature, the creation of new data files for fields of information such as phytosociology, biosystematics, etc. In addition, it would involve the creation of a database for various kinds of search. With this planning in mind, it was decided to appoint an information scientist, a bibliographer/taxonomist and a part-time secretary in late 1981. A further bibliographer/taxonomist and part-time secretary were appointed in the second half of 1982, and one more bibliographer/taxonomist was appointed in early 1983.

1. *Equipment*

In May 1982, a SYSTIME 5000 Computer System was installed, consisting of a DEC PDP 11/34A processor with 256K MOS, a Lark disc drive with 2 x 6.6 Mb discs, a RSTS/E operating system and a systems VDU. Later a further disc drive, two VDUs, a printer and a modem were added. A BBC microcomputer viewdata compatible system, consisting of the model B microcomputer, disc drive, colour monitor, acoustic coupler, PRESTEL terminal software and file transfer software was added later.

Two substantial software packages have been bought, namely ACCESS and COMPUTEX. ACCESS is an indexing system produced by SYSTIME, and will form an integral part of the final database. It allows fast response times for on-line searches which otherwise would be difficult to implement. COMPUTEX is a viewdata system which allows output to be presented in seven different colours, as in PRESTEL, and allows electronic messages to be sent and received from a remote source.

To obtain full benefit from the system, a viewdata compatible peripheral should be used. This may simply be a viewdata television set, or, if down-loading of information is required for further processing, then a viewdata compatible system such as the BBC microcomputer with colour monitor and PRESTEL terminal software should be bought. If information only needs to be stored, however, then an audio-cassette tape dump can be made from the back of the viewdata terminal, or colour slides may be prepared from photographs of the terminal screen. It is also envisaged that printed output (either seven colour or black and white) can be sent to users on demand.

2. Implementation

(a) Phase I. To ensure that the system adopted will meet, as far as possible, the diverse demands likely to be made on it by a variety of research users, and to avoid duplication of effort and resources, the first phase of the project has involved the analysis and review of existing relevant documentation systems. To this end close liaison has been established with the Mediterranean Check-list project (Greuter *et al.*, 1981) and visits have been made to its three centres, the Botanischer Garten and Botanisches Museum, Berlin, the Conservatoire et Jardin botaniques, Geneva and the Ecothèque Méditeranéenne du CNRS, Montpellier. Likewise, useful discussions have been held with the staff on the IUCN Threatened Plants Committee secretariat at Kew and common strategies planned in certain areas of mutual interest. Links have also been established with the Vicieae Database Project at Southampton University under the direction of Dr F. A. Bisby (see Adey *et al.*, Chapter 15 this volume). Discussions have been held with several of the major abstracting and information services such as Biosciences Information Service, Philadelphia and Biosis (UK) Ltd, Boston Spa, the Commonwealth

Agricultural Bureaux, Farnham Royal, Slough, and it is hoped to establish a working relationship with the CNRS Bulletin Signalétique. There is a close liaison with the AGRIMED "Flora of Mediterranean Weeds" pilot project at Reading and contacts have been initiated with the Secretariat of the Committee for Mapping the Flora of Europe, Helsinki (Jalas *et al*., 1972–80). In addition to the immense help resulting from these and other visits and consultations, further views on potential consumer demands were based on a survey of participants at a regional meeting of the Linnean Society of London and an international conference of the Systematics Association, both held at Reading in 1982.

In order that the consultations already initiated may be continued and extended, a Panel of Potential Users has been inaugurated. This currently has 12 members, mostly in the British Isles, but it is being expanded to include colleagues from continental Europe. The members of the Panel will appraise alternative formats of output information sent to them and, as several have already done, suggest what additional classes of information might be incorporated into the database. The greatest importance is attached to such feedback in ensuring that the Project meets the real needs of the consumers for whom it is intended in the most economical manner possible.

Alongside the consultations and other steps just referred to, Phase I of the Project has proceeded to lay the foundations of the database by incorporating relevant information from the published volumes of *Flora Europaea*. The procedure, approved by the Steering Committee at its meeting in 1982, involves the incorporation of the names of all taxa, their authorities and literature citations, mentioned in *Flora Europaea*, in such a way that they can serve a variety of user-application enquirics. Colour coding will be used both in on-line video displays and in some paper print-outs to differentiate accepted (white), passim (mauve), synonymous (green) and doubtful (red) names; post-*Flora Europaea* names appear in yellow. Geographical information can be searchable by country or by groups of countries; free text notes on the geographical distribution of taxa as used in *Flora Europaea* can also be added. Chromosome numbers given in *Flora Europaea* are being entered, together with supporting references taken from the *Flora Europaea Checklist and Chromosome Index* (Moore, 1982). It is also planned to incorporate further information from *Flora Europaea*, such as the bibliographical appendices.

A complex and time-consuming process has been the incorporation into the database of the synonyms which are given only in the index in *Flora Europaea*. The species (or subspecies and genus) in the text to which they are referable are indicated in the index by page number and, sometimes, taxon number in parentheses.

It is planned to hold a four-day workshop in Vienna, in 1983, to allow the Steering Committee and representatives of major European taxonomic centres to discuss the systems and assess its viability in terms of their needs.

(b) Phase II. Although most of the information mentioned in the description of Phase I of the project has been entered into the computer, these computer files will not be built into a database structure until the additional data fields it is hoped to cover have been defined. Likely additional fields are:

(1) *Additional references:* It is anticipated that the bibliographical files will be open-ended that the basic core of taxonomic and chromosome references will be extended by continuous additions in all fields.

(2) *Illustrations:* It has been requested by members of the User's Panel that a file of illustrations to European plants should be set up. The use of the archive material in the Botany Library of the British Museum (Natural History) has already been offered as a basis for this work.

(3) *Ecology:* The ecological information given in *Flora Europaea* is somewhat scant, but from a study made from a 5% sample of the ecological entries in the Flora, it is obvious that some standardization of ecological terms under headings such as "local conditions", "substrates", "plant preferences", "site-types", etc. is desirable, but not easily obtainable.

(4) *Weedy plants:* Owing to a close liaison having been established with the feasibility study for the projected "Flora of Mediterranean Weeds" of the AGRIMED Programme, it has been possible to prepare a joint list of possible field headings on this subject.

(5) *Conservation status:* After talks with staff of the IUCN Threatened Plants Committee Secretariat at Kew, it was proposed that information on the conservation of European plants should be presented in this project in conjunction with ecological and phytosociological data.

(6) *Phytosociology:* At this stage it is assumed that the presentation of phytosociological data to at least the level of the alliance

will be based on a "Classification of Plant Communities for Conservation Purposes" by Professor E. Dahl (unpublished) which is being followed by IUCN, and, to facilitate its use, an index to this document has been prepared at Reading. By this means it is hoped that the links with conservation status will be emphasized and that this new standard for European data will be accepted and endorsed.

(7) *Biosystematics:* This area will necessarily contain many fields, the most important of which will be cytology, genetics, genecology, population variation, breeding behaviour and reproductive capacity.

(8) *Phytochemistry:* This subject provides a wide range of characters which are important in such fields as plant classification, plant–animal relations and ecology. A screening of such standard works as Hegnauer (1962–69) and Gibbs (1974) may provide information which should be able to be incorporated into the database.

(9) *Pollination:* The relevant data in this subject are the plant characteristics which may play a role in pollination by animals, and also the names of pollinators. A good start in this area could be the pollinator file which is available at Leiden University.

3. The European Documentation System and the User

One of the main aims of the project is to cater for different types of user. For this reason, programs will be written to allow users to search the database even if they have had no previous computing experience. Obviously, there are practical and financial limits to the amount of searching which can be done on-line from a remote source, and principally for this reason there will be a message facility which will allow a user to type his request on to his screen, and send it to the computer. The computer will then be able to inform a member of staff that there is a message for him from the user and can relay that message on request. Once the requested search has been done, a message can be sent back to the user that his information is on a numbered "page" on the system. Thus, the next time the user connects up to the system, he can read the answer to his search by typing in the "page number" given to him in his message. The message facility may also be used by workers to inform the database staff of amendments and corrections, thus avoiding the delays involved in postal circulation.

It is not intended to give any judgement on the value of any updating or amendment of taxonomic or nomenclatural published information after *Flora Europaea* but rather to present it in a colour coding which indicates only that the material is of post-*Flora Europaea* origin. It will be left to the user to decide whether to accept or reject updated or amended material. It is however, anticipated that taxonomic revisions of particular groups will be commissioned at a later stage, as and when funds are made available, and the database will be modified to take these revisions into account.

It is intended to produce documents from the information held on the computer from time to time. These will consist of draft surveys to test consumer reaction, biosystematic surveys to provide whatever biosystematic information is available on certain taxonomic groups and checklists of various kinds which are of particular interest to any group of users. It is also intended to produce an updated synonymic checklist of European flowering plants and ferns with distributional information, as soon as possible.

In summary, the system is being planned in such a way as to provide users of taxonomy with a relatively simple means of obtaining reasonably up-to-date information on European plants from a central source, presented in a readily comprehensible way similar to that used by existing commercial page-text viewdata systems. It is hoped that the programme, if successful, will become at least partly self-supporting after the completion of the 5 years during which it is part of the Additional Activity in Taxonomy.

REFERENCES

ESF (1977). "Taxonomy in Europe". [ESRC Review No. 13.] European Science Foundation, Strasbourg.

Gibbs, R. D. (1974). "Chemotaxonomy of Flowering Plants". McGill-Queen's University Press, Montreal.

Greuter, W., Burdet, H. M. and Long, G. (eds) (1981). "Med-Checklist: I, Pteridophyta". OPTIMA, Geneva and Berlin.

Hegnauer, R. (1962–69). "Chemotaxonomie der Pflanzen", 5 vols. Birkhauser, Basel.

Heywood, V. H. (1957). A proposed Flora of Europe. *Taxon* 6, 33–42.

Heywood, V. H. (1958). Flora Europaea – a progress report. *Taxon* 7, 73–79.

Heywood, V. H. (1962). The information problem in biosystematics. *Nature, Lond.* 193, 935–936.

Heywood, V. H. (1978). European floristics: past, present and future. *In* "Essays in Plant Taxonomy" (H. E. Street, ed.), pp. 275–289. Academic Press, London and Orlando.

Jalas, J. and Suominen, J. (1972–80). "Atlas Florae Europaeae", 5 vols. Helsinki.

Moore, D. M. (ed.) (1982). "Flora Europaea Check-list and Chromosome Index". Cambridge University Press, Cambridge.

Solbrig, O. T. and Gadella, T. W. J. (eds) (1970). "Biosystematic Literature". [*Regnum Vegetabile*, 69.] International Association for Plant Taxonomy, Utrecht.

Tutin, T. G., Heywood, V. H., Burges, N. A., Moore, D. M., Valentine, D. H., Walters, S. M. and Webb, D. A. (eds) (1964–80). "Flora Europaea", 5 vols. Cambridge University Press, Cambridge.

Webb, D. A. (1978). Flora Europaea – a retrospect. *Taxon* 27, 3–14.

8 The Database of the IUCN Conservation Monitoring Centre

D. C. MACKINDER

IUCN Conservation Monitoring Centre, c/o The Herbarium, Royal Botanic Gardens, Kew, Surrey TW9 3AB, UK

Abstract: The IUCN Conservation Monitoring Centre's database is split into four main units united by a common geographical and taxonomic skeleton. The four units deal with threatened animals, threatened plants, protected areas and wildlife trade. The purpose of the database is to produce various outputs synthesizing data relevant to any given conservation problem from all four units of the database. Information is stored in the computer in two main forms; in data files, which hold data in a rigid format, and text files which permit a more flexible format allowing for the variability of biological data. Data files are used to store summary information such as data on distribution and conservation status, as well as holding pointers to the text files which expand on the coded information. Geographical data is stored using codes which may be translated into political or biogeographic area names as required. Taxonomic information is held in a tree structure which permits any number of taxonomic levels and enables any taxon to be defined in a number of different ways, thus catering for taxonomic differences of opinion.

INTRODUCTION

IUCN (the International Union for Conservation of Nature and Natural Resources) set up the IUCN Conservation Monitoring Centre (CMC) in 1981. CMC is composed of four units, the Threatened Plants Unit, the Species Conservation Monitoring Unit, the Protected Areas Data Unit and the Wildlife Trade Monitoring Unit. The Centre has the task of providing information on the status of conser-

Systematics Association Special Volume No. 26, "Databases in Systematics", edited by R. Allkin and F. A. Bisby, 1984, pp. 91–102. Academic Press, London and Orlando.

DATABASES IN SYSTEMATICS
ISBN 0 12 053040 6

vation worldwide to IUCN, the World Wildlife Fund, and to the Global Environment Monitoring System run by the United Nations Environment Programme, as well as to the wider conservation movement. This information is provided in the form Red Data Books consisting of data sheets on threatened plants and animals, Protected Areas Directories consisting of data sheets on national parks and protected areas, scientific articles, special reports and other tabulations of conservation information; such information is essential to the success of the conservation movement. IUCN's conservation objectives are succinctly stated in the World Conservation Strategy (IUCN, 1980) and have played a large role in shaping the CMC database.

A major factor behind the structure of the CMC database is that it must remain flexible so that it can match the changing needs of conservation. The detailed structure of the data files reflects the compromise between the data that CMC would ideally like to have and that which it is practical to collect, code and use at present. Thus, for others whose interests overlap those of CMC, in certain areas the CMC database will undoubtedly not go far enough or will over-simplify, especially since the computer side of the database is still fairly new. However, it should be remembered that this database is continually developing by becoming more comprehensive and covering a broader scope, so these problems should decrease. The current structure of the data files is not seen as permanent or as a constraint to the expansion of the database, indeed the number of different pieces of information and the amount of detail coded is continually increasing.

DATA STORAGE METHODS

Data is stored in two main ways, in summary form in data-processing files and as text in word-processing files. Storing data in summary form imposes restrictions on exactly what data is stored since not all pieces of information may be coded easily and efficiently. Storing data in this form, however, allows one to exploit the power of computers first to extract all information relevant to any given problem and, secondly, to sort it into any desired order for presentation; this is (in computing jargon) data processing or DP. To avoid the problems of forcing inherently variable biological information into a DP format, information is also stored in word-processing (WP) documents which allow a complete coverage of the more complex details relating to any given piece of information. However, as a result of this freedom it is much more difficult to programme a computer to extract and rearrange the information stored in text files; for instance, it would be possible to extract all paragraphs which contained the word Peru from all word-processing documents, but then simply printing them end to end would probably produce an unconnected sequence of paragraphs. By storing the data in these two different forms (DP and WP) it is possible to get the best of both the rigid DP and flexible WP environments.

Storing information both in DP and WP formats is particularly easy for CMC since both functions are performed on the same computer (a Wang VS80 installed in the Centre's Kew offices). The Wang allows easy transfer of information between DP and WP and vice versa. CMC also uses the smaller Wang OIS-115 in its Cambridge offices for word-processing and in the future to provide (via a telecommunications link) access to the DP facilities of the Wang VS. Most outputs from DP are in the form of word processing documents, which allows last-minute qualifications (which could not be handled by the more rigid DP system) to be added before printing the results, as well as allowing the output to be passed directly into typesetting equipment to produce high quality presentations. One of the pieces of information stored in DP is the name of the WP document containing the text giving a fuller description of the information coded in DP, thus making possible the automatic selection (and possibly printing) of all data sheets relevant to any given problem.

THE DATABASE UNITS

The CMC database is split into four main units – animals, plants, protected areas and wildlife trade, which are united by a common skeleton of geographical and taxonomic data. The files in each of the units are designed both to fulfil the needs of that section and also to allow integrated outputs to be produced from the whole database.

1. The Geographical Skeleton

One of the major criteria for selecting and sorting data from the database is the geographical information. Therefore, a standard, flexible system for coding geographic data has been devised. The fundamental design goal has been to devise a scheme which is the lowest common denominator between the available alternatives for classifying geographical areas. Having done this, it is possible to programme the computer to produce outputs according to various different geographical classifications, for instance sometimes the areas of interest are biogeographical whereas on other occasions they are political. The areas which comprise the CMC area classification system are defined such that each one is not, if possible, split into more than one political or biogeographical area. Therefore, these areas may be translated directly into political or biogeographical areas for use in outputs. Devising an area classification in this way has the drawback that sometimes the areas will be defined on a much finer scale than the available geographical data to be coded. Distribution data are often very vague and cannot be coded easily using a very fine-grained system, without giving the data a misleading appearance of accuracy. Thus the CMC area system also includes certain larger areas, for instance, continents, to counter this problem. For large countries, such as the United States, the basic CMC areas are the individual states, the United States as a whole is also included for cases where distribution to State level is not known. Similarly, countries made up of islands (e.g. Indonesia) are split into the individual islands. Thus the CMC area classification system includes areas at three broad levels; continents, countries and then states, provinces or islands within larger countries.

Geographical information is coded using CMC area codes in all database files, therefore it is a trivial matter to use different sets of code translation tables to produce area names in political or biogeographical terms, either in English or any other language. CMC currently has an almost complete set of French area names and has started to produce German and Spanish translation tables. It is also possible to translate CMC area codes into the International Organization for Standardization (ISO) codes and then produce output in terms of ISO country names.

2. *The Taxonomic Skeleton*

Scientists use the system of binomial Latin names so they may refer unambiguously to the taxa with which they are concerned. It is therefore vital that those names are spelt correctly and it is virtually essential that a database using taxonomic names have some method of ensuring that all database outputs use the correct taxonomic names. Thus CMC has built up various files containing the majority of taxonomic names that CMC requires. These files have other important advantages besides allowing taxon names to be checked. Computers are able to store numbers much more compactly than characters; for example, assuming that it takes 8 binary bits to store one character, a binary number between 0 and 65535 may be stored in the same amount of space as two characters. Thus code numbers for each name may be stored instead of the actual names, thereby saving a considerable amount of space especially when there are many records in the database for the same taxon, as is the case in the trade database files. Look-up files allow this coding to be done automatically whilst simultaneously checking the spelling of the names being coded. If numerical codes are stored instead of taxonomic names, the spelling of a name may be changed throughout the database by making one alteration to the name look-up file, rather than by searching for every occurrence of that name in the database. The methods CMC has used to construct these taxonomic look-up files have varied for the plant and animal sections due to the different taxonomic problems involved.

The Threatened Plants Unit (TPU) uses three taxonomic levels above species – genus, family and division. In cooperation with the staff at the Royal Botanic Gardens at Kew, the TPU have created data files containing all generic and family names accepted by the Herbarium at Kew. All families within a given division are given consecutive numbers, enabling genus, family and division information to be coded by a family number and a genus number. At present specific and subspecific names are not coded in the TPU files since each name would not occur many times and, therefore, the space saving does not offset the extra work needed to create and maintain the look-up table files; these names are simply stored as characters.

For animals (especially invertebrates) however, many more taxonomic levels are needed. Thus any one taxon must be coded as belonging to several taxa at higher taxonomic levels. One solution is to devise a taxon number composed of several consecutive sets of digits each of which describe successively lower taxa. The number could start with one digit describing the kingdom, two digits for the phylum within the kingdom and so on. The ISIS (International

Species Inventory System) described by Seal and Makey (1974) is such a system. An advantage is that taxonomic sorting is very easy, as is the process of determining the taxonomic name for any given code. Making taxonomic changes is the problem since, for example, to move a genus between families the code numbers for the genus and all its submembers must be altered in both the look-up tables and in the database itself. In the case of a large database this may result in altering many files. The cause of this problem is that two separate pieces of information are being coded in the same number; the taxon's taxonomic position and the taxon's individual identity. These two things are separate, since the biological information relating to a taxon (its individual identity) does not necessarily change if its taxonomic position is changed. A solution is to use arbitrary unique numbers to hold the individual identity of the taxa and then make a file detailing the taxonomic relationships. These may be handled very flexibly by using a tree-like structure, if the simple assumption is made that any taxon may only be a submember of one other taxon. Thus every taxon has a pointer to the next higher taxon in the taxonomic tree, except for synonyms which have a pointer to the acceptable name for the taxon.

Such a "taxa-tree" may be constructed by making a file which, for every taxon, holds its name and number, the number of the taxon of which it is a submember (or synonym), a code for the taxonomic level (e.g. species, genus, etc.) and a taxonomic sorting code (Fig. 1). There is no requirement that all the submembers of any taxon be at

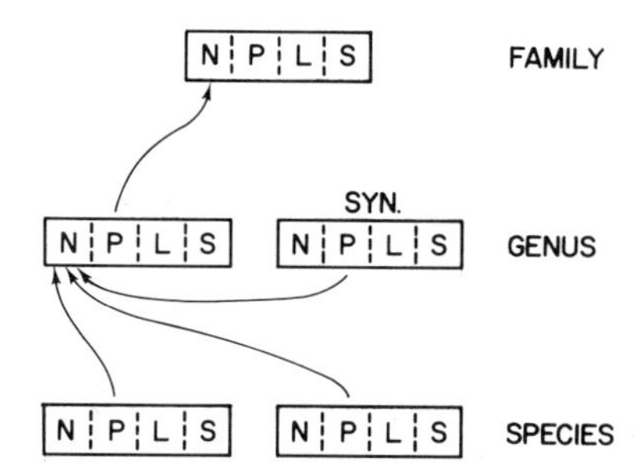

Fig. 1. Five records from the taxa-tree file illustrating the links between the records (see text for details); not all fields within the records are shown. Key: N = taxon number, P = pointer to next higher taxon (= number of next higher taxon), L = taxonomic level code, S = within-level taxonomic sorting code), SYN = record defining a synonym (a generic synonym in this example).

the same taxonomic level, or that the same number of taxonomic levels are used in all "branches" of the tree. This is, in effect, a computer representation of the tree-like structure used by taxonomists.

The sorting code allows all the submembers of any given taxon to be placed in taxonomic order for database outputs. If the sort codes are only used to sort taxa above a certain taxonomic level whilst taxa below that level are sorted alphabetically then hybrid taxonomic-alphabetical sorts may be produced (e.g. some outputs need to be ordered taxonomically down to family level and then alphabetically by genus and species name).

The rule that a taxon may only be a submember of one other taxon may be relaxed to allow alternative definitions (or "flavours") of taxonomic names. A single tree structure may then be built which contains several different alternative taxonomies. This has the advantage that the entire database may use a single set of taxon numbers even though different names or ordering of those names are to be used for different outputs from the database. The main problem with this method is that it requires considerably more programming time to set up than the ISIS-style numbering systems, however the resultant advantages in terms of taxonomic flexibility should be well worth the initial investment in programming time.

3. Plants

Lucas and Synge (1978) estimated that there were 20 000 to 25 000 plant species (about 10% of the total) threatened worldwide. With such a vast number of taxa to deal with the TPU was the first unit to require the powers of a computer to store, select, sort and display their data. The main TPU data file consits of taxonomic, distribution and conservation status data for 13 000 species and subspecies of plants so far identified as threatened. The database also includes non-threatened members of certain groups of special conservation concern, for example the cycads and tree ferns. Conservation status is coded both for individual areas, broader regions and the world as a whole. Details of the threats facing European taxa and the habitats in which they occur are now being added to this framework.

The plants data file has been used to produce regional or country lists of rare, threatened and endemic plants (e.g. IUCN Threatened Plants Committee Secretariat,1982 and in press; Leon *et al.*, in press). Similar lists of taxa from areas of particular conservation concern (e.g. the Mascarenes) have been produced for distribution by the TPUs Botanic Gardens Conservation Coordinating Body in order to collect details on threatened plants in cultivation.

4. Animals

Recently work has begun on coding the data accumulated on threatened animals by the Species Conservation Monitoring Unit. Especially for higher vertebrates, the number of taxa concerned is very much less than is the case for plants; however, there is much more information available about each taxon and, therefore, it has been both possible and necessary to start the animal database with a much more detailed coding scheme. The information about each taxon in the animal database is stored in the form of one main record, holding data which applies to the taxon worldwide and one or more distribution records detailing the areas in which the taxon occurs. The main record contains taxonomic data, conservation data and information on any text file in the WP side of the database which deals with the taxon. ISIS codes are stored for each taxon to allow compatibility with databases which use the ISIS system to code taxonomic information, and in particular to allow close links with the IUCN Environmental Law Centre (ELC) database in Bonn, but are not used for taxonomic purposes by the CMC database. The distribution records also store data relating to the taxon in each area. Part of this information consists of two links, one to the protected areas files and one to the ELC database. At present the links simply say whether more relevant information is available in the legal or protected area files for the current taxon in any given area.

To date, some 2400 taxa have been entered representing all mammals, reptiles and amphibians covered by the Red Data Books (e.g. Thornback and Jenkins, 1982; Groombridge, 1981), or on the CITES appendices or which have been covered by other CMC reports; the majority of molluscs and fish currently under consideration for the forthcoming Invertebrates and Fish Red Data Books have also been entered. Work has recently begun on coding data for birds. The initial aim is to enter records for all animal taxa of conservation concern.

5. Protected Areas

The protected areas section of the database holds summary data on some 4500 national parks and protected areas and is described more fully by Harrison (in press). This information allows an assessment of the extent to which the different biogeographic provinces of the world are covered by protected areas (e.g. Harrison *et al.*, 1982) so that recommendations about extensions to the world system of protected areas may be made. Programs exist which can select protected areas on the basis of any of the different pieces of information coded and then display details about the areas matching the selection criteria. The UN List of National Parks and Protected Areas (IUCN, 1982a) has been produced from this section of the database, as have the tabulations for the recent IUCN Directory of Neotropical Protected Areas (IUCN, 1982b). Another data file contains information on current and projected staffing levels in the parks systems of different countries and is being used to compare staffing levels and make recommendations about future staffing levels and training schemes.

6. *Wildlife Trade*

The Convention on International Trade in Endangered Species of Wild Fauna and Flora (CITES) is designed to control trade in threatened species of animals and plants by requiring both importing and exporting states to issue permits for trade in such species. All Parties to the convention are required to submit, to the CITES secretariat, annual reports detailing permits issued for trade in CITES listed species. Data from these reports forms the basis of the wildlife trade section of the database and is analysed for CITES by the Wildlife Trade Monitoring Unit.

In the case of trade between Parties to the convention there should be two records of that trade in the CITES data files, one from the importing and one from the exporting country. Analysis programmes have been written which allow the exports reported by a country to be compared with imports other countries have reported receiving from the exporting country. Similar analyses for a countries imports are also possible. Reports may also be produced for non-Party states or Party states which have not reported. This data enables an assessment of the volumes and trends in international trade in threatened species to be made, and also permits an analysis of the effectiveness of the convention either for different taxonomic groups or for different regions of the world.

FUTURE DIRECTIONS

1. Conservation Projects Database

To achieve maximum conservation gain from restricted funds, conservation organizations must plan their projects to minimize project overlap and minimize duplication of effort. Therefore, information on conservation projects which have been identified as necessary, and those which are in progress, is necessary. Such information also enables gaps in current conservation effort to be identified and possibly remedied. Therefore, CMC is starting to create a database containing information on conservation projects; the database has been designed in conjunction with IUCN/World Wildlife Fund staff in Switzerland who need similar information for their internal management of projects. The projects database will be tied to the rest of the CMC database by the standard CMC geographical and taxonomic coding schemes.

2. Address System

CMC depends for much of its data on a large international network of contacts. Currently the addresses of these contacts are handled in a variety of ways, both in the computer and in card-filing systems. Therefore CMC is planning to expand the present computerized address system to allow the taxonomic and/or geographical specialities of CMC contacts to be coded, thus enabling automatic selection of all contacts relevant to a certain problem.

3. "Area-tree'

The construction of an "area-tree" to define the relationships between the various geographical areas in the current CMC area code system will greatly increase the power and flexibility of geographical coding and analysis of conservation data within the CMC database. The area-tree will be similar in structure to the animal taxa-tree (described above).

4. Bibliography System

As the volume of scientific literature on conservation and related topics grows so does the need for a computerized method of storing bibliographic details relevant to the work of CMC. Currently, TPU in conjunction with the Library of the Royal Botanic Gardens at Kew are operating a trial version of a bibliographic information system covering references relevant to plant conservation; it is planned to expand this to cover all references used by CMC. As with projects and addresses each reference will be coded for the taxa, areas and topics to which it is relevant. In addition to allowing automatic selection of references pertinent to any conservation problem, it is hoped to be able to use this information to produce reference lists for future CMC publications automatically, without having to retype the citation details of each reference every time it is used.

CONCLUSIONS

Two of the major methods of selecting conservation information are by taxonomic and geographic data. Therefore, the CMC database needs a well developed, flexible method for coding, and later selecting and sorting taxonomic and geographical data. Construction of tree structures (detailed above) appears to be the best method to handle such coding systems, since it does not impose any restrictions on the number of levels in the classification system and new codes may be added to the system without needing to alter existing codes. It is envisaged that such tree structures will also be used to manage other types of information, such as habitat classifications, as time proceeds. This powerful and flexible structure, together with the vast amount of information available to CMC through its extensive international contact network allows the CMC database to make a major contribution to world conservation.

ACKNOWLEDGEMENTS

The author would like to acknowledge the help and advice received from the staff of CMC, especially Jerry Harrison and Jane Thornback.

REFERENCES

Groombridge, B. (1982). "The IUCN Amphibia-Reptilia Red Data Book". Part 1, Testudines, Crocodylia, Rnynchocephalia. IUCN, Gland, Switzerland.

Harrison, J. (in press). Protected areas monitoring by the International Union for Conservation of Nature and Natural Resources. *In* "Tropical Rain Forest:

Ecology and Management – Supplementary Reports" (Chadwick, A. C., Sutton, S. L. and Burch, P. R. J., eds). Leeds Philosophical and Literary Society, Leeds.

Harrison, J., Miller, K. and McNeely, J. (1982). The world coverage of protected areas: development goals and environmental needs. *Ambio* **11**, 238–245.

IUCN (1980). "World Conservation Strategy". IUCN-UNEP-WWF, Gland, Switzerland.

IUCN (1982a). "1982 United Nations List of National Parks and Protected Areas". IUCN, Gland, Switzerland.

IUCN (1982b). "Directory of Neotropical Protected Areas". IUCN, Gland, Switzerland.

IUCN Threatened Plants Committee Secretariat (1982). The rare, threatened and endemic plants of Greece. *Ann. Musei Goulandris* 5, 69–105.

IUCN Threatened Plants Committee Secretariat (in press). "List of Rare, Threatened and Endemic Plants in Europe". [Nature and Environment Series.] Council of Europe, Strasbourg.

Leon, C., Lucas, G. and Synge, H. (in press). The value of information in saving threatened Mediterranean plants. *In* "Plant Conservation in the Mediterranean Area" (Gomez-Campo, ed.).

Lucas, G. and Synge, H. (1978). "The IUCN Plant Red Data Book". IUCN, Morges, Switzerland.

Seal, U.S. & Makey, D. G. (1974). "ISIS Mammalian Taxonomic Directory". International Species Inventory System, Minnesota Zoological Garden, St. Paul. Minnesota.

Thornback, J. & Jenkins, M. (1982). "The IUCN Mammal Red Data Book". Part 1. IUCN, Gland, Switzerland.

9 | ISIS — An International Specimen Information System

N. R. FLESNESS

ISIS, 12101 Johnny Cake Ridge Road, Apple Valley, Minnesota 55124, USA

P. G. GARNATZ

Computer Software Associates Inc., P.O. Box 8089, Roseville, Minnesota 55113, USA

and

U. S. SEAL

U.S. V.A. Hospital, Minneapolis, Minnesota 55455, USA

Abstract: ISIS (the International Species Inventory System) is a nine-year-old specimen information system, with data on more than 100 000 animals held by zoological gardens and related facilities in twelve countries. ISIS provides 180 participating institutions with standard data forms, current information on all collections and some special reports available on request for specified species. The system uses standardized taxonomies, and codes for taxa and institutions. The system permits use of local specimen identification schemes, and also uses double reporting of specimen transfers. ISIS has successfully built a large multi-institutional computerized specimen catalogue.

Systematics Association Special Volume No. 26, "Databases in Systematics", edited by R. Allkin and F. A. Bisby, 1984, pp. 103–112. Academic Press, London and Orlando.

DATABASES IN SYSTEMATICS
ISBN 0 12 053040 6

INTRODUCTION

ISIS, a nine-year-old computerized specimen information system, now includes data on more than 100 000 specimens held in 180 collections in 12 countries. Multi-institutional master catalogues like ISIS can be a success, although they are not always so (Sarasan, 1981). They are valuable for locating specimens and retrieving detailed information about particular individuals and are the only feasible way to evaluate quickly and efficiently the combined strengths and weaknesses of a large group of collections.

The need for such master catalogues is greater when the collections are alive. Living collections require dynamic management to compensate for mortality. In a growing number of cases, replacements are not available from the wild (for reasons of conservation ethics, international treaty, cost and, in a few cases, extinction in the wild) and sustained reproduction and sophisticated long-term population management have become necessary.

WHAT IS THE INFORMATION NEED?

For economic and logistic reasons, the population at any one facility is often just two, three, or four – too small to have any demographic stability. Chance events often cause these local populations to become extinct, and boom/bust cycles occur for a number of reasons. There is a need for cooperative management of the entire population in captivity. This requires, at minimum, pooled information.

A second need is for pedigree data and genetic analysis. This particularly applies to captive populations, where little or no immigration can be expected in the long term. Small, closed populations lose genetic variability through drift and this has long-term consequences, as future evolutionary potential is reduced. Separately, inbreeding produces individuals that are often less viable, and may have other weaknesses (Flesness, 1977; Ralls *et al.*, 1979). With specimens that move through several institutions in their lifetime, keeping track of individual relationships is difficult for any one facility. The need for a central entity to track such relationships, and also analyse population-level genetic factors, is apparent.

BUILDING THE DATABASE

There are several critical facts about the environment ISIS must function in, which greatly affect design:

(1) Not all facilities participate – specimens may transfer out of, or into, the group of participating institutions.
(2) The system must integrate well with existing institutional specimen record systems – which means accepting the specimen identification *they* already use.
(3) Specimens are frequenly loaned – and there is a need for information on both the legal holder and physical holder of the specimen.
(4) Many facilities start ISIS participation without good internal records – improving their internal records is a system goal.
(5) The system itself must be simple, efficient and quite economical (ISIS now has a staff of four, but half the system support comes from grant and contract sources).

To build a system subject to these constraints, a mix of high technology, and deliberately low technology, has been used. The system was designed and built in 1973 (Seal and Makey, 1975). Recently, the database portion has been radically altered – but the ISIS system, of which the database is a component, remains largely the same.

An unusual feature of the ISIS system is the method of data capture. To solve simultaneously the problem of providing a good internal records system for individual institutions, and the problem of economically capturing data for entry into ISIS, a single universal specimen record was created. At present this is a paper record, which is provided to participating institutions by ISIS. Fifty thousand archive-quality, two-part forms are distributed by ISIS, filled in by 180 records persons in twelve countries, and one copy of the form returned to ISIS. The other copy forms the base for a good internal records system for each institution.

With mortal and mobile specimens, the basic record is a transaction – a record of change. Each transaction represents a birth or death, sale or purchase, loan out or loan in. Each such event is reported to ISIS in a standardized format, ready to be keyed into a machine.

There are a large number of animal registry systems, for special breeds of domesticated stock (Boucher, 1980). ISIS differs from these systems in two conceptually important ways, by using double reporting, and by not using a single lifetime ID (specimen identification).

Because not all facilities in the world participate in ISIS, we cannot use a single transaction, as the breed registries do, to move a a specimen from one facility to another in the database. Without universal participation, failure of a recipient to forward a single transfer form would cause database inaccuracy. Instead, ISIS uses a double reporting scheme. The first institution reports to ISIS on shipping the specimen to the second, and the second institution usually reports receiving it from the first. This gives each institution control over the data representing their collection, and works within limits even when one of the two is a non-participant. It does, however, make a more complex database, since the chronology of reporting may mean that ISIS receives word of arrival before word of departure, the database must temporarily permit the specimen to be in two places at once.

The other unique approach used by the ISIS system concerns specimen identification (ID). The average specimen may visit two or three facilities in its lifetime; long-lived species may transfer up to ten times. To make it straightforward for institutions to participate, we let each use its own specimen identification scheme. For database uniqueness, we simply add the institution code and taxonomic code to their specimen ID to assure that no two specimens will have the same database key.

One specimen may therefore acquire a string of IDs – one for each institution visited. The ISIS system is designed nonetheless to permit tracking of one individual from birth to death. This is achieved by requiring each succcessive institution to report what the previous institution used for a specimen ID. The database links all the data for one specimen together. The string of specimen identifications does not in itself create database problems; the problems that arise do so because of transfer through a non-participating institution.

STANDARDIZATION

Because of the requirements of capturing data in machine-keyable form directly from 180 individual records persons, considerable standardization has been adopted. Two critical data items about the specimens are their taxon, and their holding institution. Each may have multiple names for one entity, i.e. synonymy, and dynamic nomenclature, taxa rise and fall, institutions come and go. Taxa present one additional problem when compared to institutional identifications, as taxa can be declared never to have existed. Use of a standardization institutional name is essential for consistent use of the collected data, as institutions are often referred to by a formal name, the city of their location, and at least one acronym. Thus the Henry Doorly Zoo of Omaha, Nebraska is referred to as HDZ as well as "Omaha", and "Henry Doorly". The problems of synonymy in taxonomic nomenclature are well known.

The ISIS solution is to provide directories which contain standardized taxonomies of institutions, geography and animals. These directories are prepared and maintained using other compilations and relevant literature, and are updated as needed. They are *not* intended as authoritative taxonomic works, but rather as a useful standard naming convention for the purpose of collecting data.

At present, ISIS directories contain about 30 000 standard entries. The ISIS World Geographic and Zoological Institution Directory contains about 3000 geopolitical entries, and about 3000 captive wild-animal facilities worldwide. ISIS Mammalian and Avian Taxonomic Directories contain about 23 000 entries, covering the world's species of these classes, plus subspecies where significant demand and opinion of ISIS users warrant them. An additional 10 000 standard taxonomic entries are forthcoming as ISIS is near to publishing Reptile and Amphibian Directories, and then collecting information on living captive specimens.

The directory approach has been effective in standardizing the taxonomy used in the ISIS database. Such standardization permits creation of reports which are known to include *all* the individual specimens in a specified taxon or institution; none could be missed because they were entered under some unrecognized synonym. This approach requires establishing a pragmatic naming consensus before collecting data.

The initial effort involved in such directories is high, while the maintenance effort is moderate. Thus far, only single entries have been used; no synonyms are included. Synonym capability would be desirable but would add to the cost.

CODING

Separate from the standardization issue is one of coding. The ISIS system uses codes for two principle data items, taxon and institution. The taxonomic directories and the geographic/institution directory are used by ISIS to provide codes for these, as well as to standardize taxonomic and institutional identification. Coding schemes for both were developed by ISIS (Seal and Makey, 1974; Seal *et al.*, 1974), and both use hierarchical codes which carry useful information internally. The geographic/institution codes carry with them the geopolitical area referred to, specified as to continent, country and next lowest political subdivision – province, state, etc. The taxonomic codes carry the specifications for the complete higher category taxonomic assignment of the taxon.

Such internal codes have advantages and disadvantages. They were intended to simplify the task of application program development. They succeed in this – it is a simple matter to select, sort, and search for geographic and taxonomic units, since the database keys carry this information directly.

These information-carrying codes have a major advantage for ISIS. Because they permit simpler database design, they enable lower maintenance and easier program development. These pay-offs are important to ISIS, which has yet to employ its first staff programmer.

The general drawback of coding, excessive clerical burden, is well known. There are also specific drawbacks to *meaningful* codes, particularly when the information being coded can and does change in ways beyond our control.

A geopolitical coding system has to be dynamic. Countries, and major subdivisions of countries, change names, are submerged into other entities, or split off from previously listed entities, with significant frequency. The choice is between modifying all database entries as change occurs, or retaining archaic geopolitical entities and codes in the database. The former option represents a higher maintenance burden, the latter option makes a search based on origin much more difficult.

More interesting are the issues about the dynamics of taxonomic coding. The problem could be called the Linnaean Dilemma. Linnaeus. after all, was the one who partly coded taxonomic classification using binomials. This idea, that the first word of the binomial indicate the genus name, is intellectually convenient over the short term. It makes it simpler to remember species affinities than if the name carried no such information. But it is precisely this that has caused the problem of published literature on one taxon being scattered among several different names as the classification of the species has changed over time.

The codes that ISIS have used to identify taxa carry the advantages, and the disadvantages, of Linnaeus' method, and carry them much further. ISIS taxonomic codes are a name, and also contain the complete classification of the organism, specifying Kingdom, Phylum, Class, Order, Family, Genus, Species (and even subspecies). Thus the name and the classification are linked much more extensively than in the Linnaean binomial.

This makes it easy to extract information on higher groups but it also means that taxonomic revision changes the taxonomic code. Since the codes also were assigned alphabetically within categories, a change at a higher rank, giving it a new number, breaks down the alphabetical order. Database entries and keys must be changed as well. In practice, nine years of operation have not encountered this problem very often, but it will increase over time.

Should future experience show that it would work to collect information with the Latin taxon name as the taxonomic identifier, ISIS might eventually "decouple" the numeric code from the taxonomic classification, use the numeric code for database purposes only, and construct input/output software so that incoming data and outgoing reports used the Latin name only. One would then build some kind of hierarchical auxiliary file to indicate the current classification and relationship of each taxon.

RECLASSIFYING THE ASHES

A second, historical problem arises because of ISIS's role as a keeper of pedigree data on living specimens. When taxa are created, or "sunk", data on some animals' ancestors becomes problematical. If ISIS has data showing that the pedigree is made up of species X, which is "sunk" into species Y, it is straightforward to change the ancestral records to Y. However if later science finds that species W was really two species, it is hard to reclassify the ancestors of today's population(s) – many of them will have been incinerated at death, for example.

HARDWARE AND SOFTWARE DETAILS

ISIS began operations on a set of IBM 370 computers, using the TOTAL database management system, and a set of COBOL applications programs. The computer centre used provided all development and operating support; an appropriate choice for a system with a staff of two. The database structure used was a typical inventory system and performed inventory tasks adequately, but could do little else.

After several years, as ISIS grew to a staff of four, it became important to create a more powerful and responsive system, which would be under the control of ISIS staff. The computer centre at the University of Minnesota was chosen, consisting of Control Data Cyber 730, Cyber 74, Cray 1B, VAX 11/780, plus microfiche and laser-print capabilities.

The new ISIS database consists of fixed-format records, one per specimen per visit to an institution. These records are indexed by database keys on taxon and institution, which also carry information on ID, date in, date out, and whether the visit was a loan arrangement. The key entries are arranged in virtual sequence access structure. A specimen's lifetime information is a string of these fixed-format records (the string bifurcates into legal and physical paths when a loan occurs).

This architecture has been built in FORTRAN (ANSI X3.9, 1978; ISO 1539-1980), as no database management system available at our host site could meet our needs economically. The structure is more powerful and flexible than the previous system, and costs about one-third as much to operate.

By far the most challenging task involved in the creation of this new database was the editing program. Nine year's previous experience with a third of a million transactions had shown the need for a sophisticated set of checks. Besides the usual validation of data items, checks were needed for consistency with previous data submitted by the same institution, and consistency with data submitted by facilities which held the specimen earlier. The program now assures data integrity with a set of checks which could only be developed on the basis of experience; for example, we now check that the female parent was alive on the date when her offspring was born, but allow for exceptions for species that hatch, and for transplanted ova.

ISIS AS AN INFORMATION RESOURCE

Despite the challenges, the ISIS system has been reasonably successful. Information on more than 100 000 specimens of mammals and birds has been compiled from more than 180 institutions in twelve countries. The database includes specimens in 2 classes, 45 orders,

198 families, 1000 genera, and 2204 species. Additionally, specimens have been entered under 950 user-requested subspecies.

In total, a quarter of a million acquisitions and removals are in the database. Of these, over 71 000 are births and over 47 000 are deaths. An additional 72 000 acquisitions and 43 000 removals represent transfer of specimens between institutions, with 8000 of these transfers representing specimen loans.

With this information, ISIS provides all participant institutions with three kinds of reports:

(1) Pooled collection inventory – at six-month intervals, age, sex, and holding institution of each specimen is reported for all taxa, along with population statistics. This report covers 60 000 specimens, is 7000 pages long and is distributed on microfiche. It is the most powerful and most widely used ISIS service.

(2) Detailed reports on a requested species – on demand, ISIS provides pedigree, demographic information, reproductive histories, etc., for a chosen taxon. These reports are increasingly used for management and research.

(3) Institution inventory and inventory-change reports are routinely provided to assist with local collection inventory needs, and to provide another check on the data in the ISIS database.

FUTURE DIRECTIONS

In the short-term, ISIS will be expanding its analytical programs to exploit the greatly increased technical capability of the new database. There is a need for routine and convenient analysis of the demography and genetics of captive populations, which ISIS will meet by developing appropriate applications software.

Over a little longer period, an important change of another sort is expected. ISIS has existed to date as a batch-oriented, remote information system. However, many of the institutions we deal with are now in the throes of building internal taxonomic databases on small, local computers. To deal with local systems, we have to accept communications media other than paper forms, such as magnetic tapes, floppy discs and telecommunications.

Also, the program is increasingly international. We are exploring the use of satellite-based digital communications links to connect small regional machines with the central ISIS database.

ISIS must also plan for distributed processing not only for data capture, but for data dissemination. There is increasingly expressed need by our users for an interactive ISIS, with telecommunications access, so that they can interrogate the database themselves. The only barrier to this is cost: as on-line disc storage (over 120 megabytes), maintenance of security and accounting systems, and development and maintenance of the query software would all be quite costly. Preliminary estimates indicate a rise of about one-third to one-half in the overall ISIS budget would be necessary to offer basic interactive services. At some future point, our user community may feel that this is worthwhile.

A general direction for ISIS development is to improve ease of use while pursuing other goals. Thus we eventually hope to reduce the degree of coding now necessary. The no frills approach taken thus far has probably been a key ingredient in the nine-year survival of the program. Relatively "low-tech" systems seem to survive the inevitable lean years better than sophisticated systems which depend critically on extensive technical staffs and accompanying large budgets.

ACKNOWLEDGEMENTS

We thank the Am. Assoc. of Zoological Parks and Aquariums, the Am. Assoc. of Zoo Veterinarians, and the Inst. of Museum Services of the US Dept. of Education for support, and Judith Block and Richard Holtzman for useful criticism and discussion.

REFERENCES

Boucher, H. (1980). "National Society of Live Stock Record Associations Directory". West Plains, Missouri.

Flesness, N. (1977). Gene pool conservation and computer analysis. *Int. Zool. Yb.* **17**, 77–81.

Ralls, K., Brugger, K. and Ballou, J. (1979). Inbreeding and juvenile mortality in small populations of ungulates. *Science, N.Y.* **206**, 1101–1103.

Sarasan, L. (1981). Why Museum Computer Projects Fail. *Museum News* **59**, 40–49.

Seal, U. S. and D. G. Makey (1974). "Mammalian Taxonomic Directory". ISIS, Minnesota Zoological Garden, USA.

Seal, U. S., Makey, D. G. and Murtfeldt, L. E. (1974). "World Geographic and Zoological Institution Directory". ISIS, Minnesota Zoological Garden, USA.

Seal, U. S. and Makey, D. G. (1975). Computer usage for total animal and endangered species inventory systems: a specific proposal. *In* "Research in Zoos and Aquariums", pp. 178–190. National Academy of Sciences, Washington.

10 The Network of Databanks for the Italian Flora and Vegetation

P. L. NIMIS and E. FEOLI

Istituto Botanico Cas. Università,
34100 Trieste, Italy

and

S. PIGNATTI

Istituto di Botanica Città Universitaria,
00100 Rome, Italy

Abstract: During the last 15 years a research group at the Botanical Institute of Trieste University has worked on the creation of a network of databanks concerning the Italian flora and vegetation. The following banks had been constructed by the end of 1982: (1) A central databank on the Italian flora, whose principal aim is to provide a standardized nomenclatural and coding system for the local banks. This bank is also useful for the solution of geobotanical and phytogeographical problems. (2) A central databank for storage and retrieval of phytosociological relevés. The aim of this bank is to provide a tool for the elaboration of synthetic monographs for the Italian vegetation. (3) A series of local databanks for floristic data, based on *ad hoc* software. These banks have been designed to store data obtained in the Floristic Cartography Project of Central Europe and may also be used in the computerization of herbaria.

The Trieste group is now working towards the creation of a network of local databanks in the form of a distributed database. The local banks will be connected with one another and the whole system will in turn be connected with the central databank for the Italian flora.

Systematics Association Special Volume No. 26, "Databases in Systematics", edited by R. Allkin and F. A. Bisby, 1984, pp. 113–124. Academic Press, London and Orlando.

DATABASES IN SYSTEMATICS
ISBN 0 12 053040 6

INTRODUCTION

The creation of efficient systems for the storage and retrieval of large data sets has always been an important element in the activity of taxonomists and phytogeographers. In recent years traditional methods have often been abandoned in favour of computerized databanks. Many such banks have been constructed in different countries (Brenan *et al.*, 1971) by researchers whose aims were often very similar. In most cases, however, the software adopted is different as, for example, in the handling of nomenclature and the coding mechanisms used. As a consequence, the possibility of information flow between different databanks in different countries, and even within the same country, is very much reduced. The coordination of different banks in which similar data are stored will constitute one of the main problems in this field for the near future.

In planning and implementing a network of computerized banks for phytogeographical and phytosociological data within the Italian territory, the authors tried to design an integrated system in which information flow between databanks plays a major role. This required a great amount of work centred on two main points: standardization of nomenclature and coding procedures, and the creation of standard software. In the following, we shall present a brief outline of the principal databanking achievements in this field within Italy since 1968.

THE TRIESTE GROUP

Our activity in databanking coincided with the foundation of the Working Group for Data Processing within the framework of the International Association for Vegetation Science in 1969. This was preceded by the presentation of a discussion paper in which Pignatti *et al.* (1968) considered the use of electronic computers for storage and elaboration of phytosociological relevés. On the suggestion of Pignatti, Cristofolini studied the storage system with visual punched cards as developed by Ellenberg and Cristofolini (1964). The Trieste team was entrusted with the problem of coding plant names for computer use and by the time of the first conference a discussion draft was presented (Cristofolini *et al.*, 1969). The final results are comprehensively treated by Pignatti (1976); a 7-digit numerical code was proposed, known as the "Trieste Code". Interestingly, an identical coding procedure was elaborated independently in the same period for the computerization of the South African National Herbarium in Pretoria.

The Trieste team prepared a first selection of 1507 relevés of salt-marsh vegetation that was stored both on punched cards and magnetic tape. The selection was distributed to six centres involved in the Working Group's activity (Kortekaas *et al.*, 1976) and a first synthetic treatment of salt-marsh data was presented by Lausi and Feoli (1979). During this first phase, which was concluded in 1975, the Trieste team was also active in designing new software for the numerical treatment of vegetation data (Feoli, 1977; Feoli and Lagonegro, 1979; Lagonegro and Feoli, 1979). The problem of coding focused our attention on the absolute necessity of standardization in nomenclature as a pre-requisite of international cooperation in the field of numerical vegetation analysis.

The Trieste team started a second phase in 1976, in a project sponsored by the Italian Research Council which aimed to construct a prototype databank for the Italian flora and vegetation. At the beginning the Trieste team was entrusted with the "botanical" part of the work, whereas the choice and implementation of software was carried out by researchers of the Istituto per le Applicazioni del Calcolo of the C.N.R. in Rome. The close collaboration between the two centres led to the development of a common language between botanists and "computer-people". The second phase terminated in 1981 with the creation of a national databank on the Italian flora (Pignatti, 1981; Nimis, 1981a,b), a prototype databank for the Italian vegetation (Anzaldi and Mirri, 1979) and a local databank for floristic-phytogeographical data based on new software created by researchers of the Trieste team (Lagonegro *et al.*, 1982). Of fundamental importance at this stage was the editing by Pignatti (1982) of the new *Flora of Italy* which provided the core of information to be stored in the central databank in a standardized and critically revised form. A further important point was the joining of the Trieste team with the Project for the Floristic Cartography of Central Europe (Pignatti, 1975, 1978). The local databanks have been designed to store and retrieve distributional data obtained in connection with this latter project.

The aim of our third phase, started in 1981, is to utilize the previously acquired experience for the creation of an integrated net of databanks in the form of a distributed database covering most of Italy.

THE CENTRAL DATABANK ON THE ITALIAN FLORA

The construction of the first databank on the Italian flora ran side by side with the editing of the new *Flora of Italy* by Pignatti (1982). The aim of the bank was the computerized storage of critically revised information concerning the ca. 6000 species of vascular plants that are part of the Italian flora. The species have been coded using the Trieste Code, which consists of 10 digits, 4 for the coding of genus, 3 for the coding of species and 3 for the coding of subspecies and varieties (Pignatti, 1973, 1976). To each species the following parameters have been assigned:

(1) *Distribution type:* the code consists of 2 digits for a total of 78 chorological groups. The first digit indicates broad distributional areas, the second specifies possible subdivisions within it. For example, plants occurring on the high Mediterranean mountains are coded with the number 4, followed by 1 (41) if they are present over the whole of the Mediterranean basin, or by numbers from 1 to 9 if they are restricted to certain sections of the Mediterranean area (respectively: N, E, S, W, NW, SE, NE, SE).

(2) *Life form:* the code consists of 2 digits for a total of 32 life form types, defined according to the system of Raunkiaer. Also in this case the first digit refers to a life-form type considered *sensu lato*, whereas the second digit specifies the various subtypes.

(3) *Environment type:* the code consists of 3 digits, for a total of 74 environment types. The first digit corresponds to broad environmental categories, the other two digits refer to finer subdivisions, e.g. number 6 corresponds to rocky or stony regosols, 63 to rocks, 631 to calcareous rocks. Of course, more than one environment type can be assigned to a single species.

(4) *Regional distribution:* this has been coded with a vector of 21 digits, where each digit corresponds to a definite region (political subdivisons, except for the regions of Abruzzi and Molise considered as one unit, and the Trieste province separated from Friuli).

(5) *Elevation range:* the elevation range has been coded with a vector of 30 digits, where each digit corresponds with an elevation increase of 100 m.

(6) *Position in the phytosociological system:* the code consists of 5 digits, and follows the system of phytosociological syntaxonomy. The first two digits are for the coding of the class, the third for the coding of order, the fourth of alliance and the last for the

coding of the association for which a certain species is considered to be faithful (Feoli Chiapella *et al.*, 1983).

With the exception of parameter 6, used only for a relatively small group of species, all the information mentioned has been stored for the whole of the Italian fiora. A further parameter is planned, consisting of the specification of species distribution within areas of different shape and extension to the political regions of Italy. This is needed because the political subdivision of the Italian territory rarely coincides with a subdivision based on meaningful phytogeographical and ecological criteria.

The system adopted for organization of the data is TAXIR (Estabrook and Brill, 1979). The database is completed by a series of interface programs elaborated by researchers of the Italian CNR, whose aim is the automatic correction of the data archive and the production of outputs in an easily interpretable form.

Among possible outputs from the databank we cite:

(a) percentage of a given life form type in a given environment type within a given region;
(b) variation of chorological spectra along an elevation gradient in different regions;
(c) types of environment occupied by the species of the genus y or the family x;
(d) check-list of the flora of a given area (or different check-lists according to environment types, elevation etc.).

In the present form the bank is potentially useful for two kinds of activities: applied geobotany (lists of species for given study areas) and phytogeography (elaboration of distributional and ecological data). We foresee the possibility of implementing the bank through the computerization of the keys: in this way it will be possible to obtain a potentially very great number of "floras" according to the different needs of various users (e.g. the flora of xeric environments in the Friulian plains, the flora of the alpine belt in the western Alps, etc.). One of the main products of the bank will be the publication of a monograph on the numerical phytogeography of Italy, planned for 1984. A preliminary part of this, based on distributional data for a limited set of 2600 species, has been published by Pignatti and Sauli (1976; see also Pignatti, 1980).

The central databank, however, is more a computerization of a book than a true databank. In this form, no further input of data is possible, since the information stored in the bank is limited to the information in the *Flora of Italy*. However, it assumes a fundamental importance in the creation of a co-ordinated net of local databanks covering most of the Italian territory, and this for two main reasons:

(1) The bank represents an archive in which the most relevant critically revised information on the Italian flora are stored. In this sense, we consider it to be a kind of "safe", whose information content can be modified only through the critical revision of new data eventually coming from the net of local databanks.

(2) The central bank provides the necessary standardization of parameters for the input in the local databanks (nomenclature, coding of environment-types, life forms, chorological groups, etc.).

The second point is of particular relevance. In fact, one of the main problems in the coordination of local databanks is the lack of a general reference system providing the standardization of parameters to be stored in the local banks. With the creation of the central databank on the Italian flora, and the publication of a check-list of the flora of Italy (Pignatti, 1981) this problem has been largely solved. This was an essential pre-requisite for the creation of a nationwide system of local databanks.

THE CENTRAL BANK FOR VEGETATIONAL DATA

The main source of data concerning the Italian vegetation consists of phytosociological studies performed in various parts of the country. Hence, the available data are mainly in the form of phytosociological relevés. In general, each relevé consists of a vector whose variables are species, with some additional information concerning locality, exposure, soil type, etc. The number of published relevés has risen impressively during the last 20 years, so that nowadays the situation is characterized by a great number of local studies and a definite lack of a more synthetic approach. In our opinion, one of the main tasks for vegetation science in the future will be the application of numerical methods in the treatment of large sets of relevés. As far as Italy is concerned, this will result in a series of compendia, each covering a particular kind of vegetation. In general, such a task will require us to extend the analysis to relevés also performed outside Italy, a fact that would raise even further the number of available relevés. The creation of a computerized databank for phytosociological relevés is an important step towards this more synthetic phase in Italian phytosociology.

Standardization of nomenclature is one of the main problems in the construction of such a bank. This is particularly true in Italy, where no modern Flora was available until recently so that different authors used widely different taxonomic concepts for the same taxon. This problem has been largely solved by means of the central bank for floristic data. The bank provides an updated nomenclatural system with check lists of synonyms up to the varietal level.

A prototype bank for vegetational data has been constructed on the basis of 450 beechwood relevés from Central Italy stored using the TAXIR system. Further data are being entered for halophytic and psammophytic vegetation, *Quercus*-woods, *Ostrya*-woods and arid meadows. The creation of the prototype required the solution of some software problems. Vegetational data in the form of phytosociological relevés are not directly compatible with the "flat-file" structure of the TAXIR system. The relations between the set of relevés (R), the set of species (S) and the set of the values assumed by $V = F(R,S)$ are not biunivocal, so that the whole structure cannot be represented in one "record type" only (Anzaldi and Mirri, 1980). In order to overcome this difficulty, the structure was transformed by application of the cartesian product $R \times S \times V$, that allows the description in one record type only. At the present stage of the work this solution seems to be satisfactory enough, although we do not exclude the future adoption of a hierarchical system for the organization of vegetational data. A series of interface programs was also written to facilitate automatic correction of stored data and to prepare output compatible with program packages for classification and ordination.

Possible outputs from the central bank for vegetational data include:

(a) a list of all the relevés in which a certain species, or a given set of species is present;
(b) a list of all the relevés with a given exposure, soil type, elevation, etc.;
(c) percentages of life form types or chorological groups in a given set of relevés.

Output such as (a) and (b) is in the form of a matrix that can be processed directly for classification or ordination. Furthermore, data concerning elevation, location of the relevés and environment type can be compared through a set of interface programs with the data contained in the central bank for floristic data, so that corrections or additions of new information are eventually possible in the latter.

THE LOCAL DATABANKS

The construction of a centralized databank for vegetational data is justified in that during the elaboration of vegetation compendia the researcher has to use a large number of relevés taken by other authors all over the country, and possibly even outside the country. A "regional" compendium or local vegetation study is completely different. In this case the work consists in comparing widely different vegetation types using relevés mostly performed by the author himself. In most cases, up to now, local vegetational studies have been successfully performed through the elaboration of relatively small data sets at a local level. The availability of a computerized databank is therefore not such an indispensable tool in this field.

There is, however, a further kind of data for which the availability of local databanks was felt to be a primary need. These are the distributional data obtained in the Floristic Mapping Project of Central Europe. The data take the form of species lists within quadrants of 5.5 × 6.5 km, for a total of 10 000 quadrants in Italy. The project started in 1966, and was originally restricted to Northern Italy, north of the Po river. At present, more than 1000 quadrants have been surveyed, and the project has been extended to the whole country (work is progressing in Latium, Puglia, Sicily and Sardinia). Local groups are responsible for different areas and for the elaboration of the relative data.

Software for the creation of local databanks expressly designed for the Floristic Mapping Project has been created by the Trieste team (Lagonegro *et al.*, 1982). A set of programs perform the following operations:

(1) Creation of the basic archive (program ARCBASE) – the program allows the transcription of a block of information into a temporary file in a standard format. Two kinds of records are introduced, the species record and the locality record. Fixed fields are assigned to the following information in a species record – the Trieste species and subspecies code, indication whether the species is critical or not, number of locality records, species name (according to Pignatti's Flora). The locality record information includes author of the survey, elevation, name of locality.

(2) Creation of dictionaries and their management (progams FASTDIZ and CURDIZ): program FASTDIZ extracts the information to be stored in the dictionaries from the temporary file

created by ARCBASE. Three dictionaries are constructed – one for the species, one for the localities and the last for author's names. Program CURDIZ allows the storage of further information regarding species (life-form, chorological group, phytosociological value), localities (coordinates of quadrants and subquadrants in which the locality is included) and authors (bibliographical citations, specifications of herbarium labels).

(3) Input of new data into the basic archive and recovery of information (program CURARBA): program CURARBA, with the functions ADD, DELETE, and REPLACE, keeps the basic archive up to date. The function SORT produces ordered lists (sorted alphabetically or taxonomically) of species or localities. The function SEARCH allows one to produce species lists for each locality (or quadrant), and the list of all the localities in which a given species has been recorded. Two auxiliary programs, ASZADIZ and VERIFIC, allow one to check possible differences between the names of the basic archive and those stored in the dictionaries.

(4) Program MAPPA produces distribution maps for the various species, specifying the main geographical detail of a given region; the presence of a species in a quadrant is represented by a series of letters indicating whether the record derives from a field survey, from literature data or from herbarium specimens, etc.

The new software has been used to create a databank on the flora of North-eastern Italy (Region Friuli-Venezia Giulia). For the province of Trieste, which is the smallest of the Italian political subdivisions, a further bank has been created by modification of the software to obtain smaller quadrants and, hence, more detailed distribution maps (Poldini, unpublished). With minor modifications, the software is also useful for the computerization of herbaria. A first prototype has been created for the Nimis herbarium (lichens) in Trieste. The software has been distributed to various centres involved in the Floristic Mapping Project. The herbaria of Trieste, Florence, Rome and Catania are also interested in making use of it.

TOWARDS A NETWORK OF DATABANKS

In creating the central databank for vegetational data and the local databanks, particular attention has been devoted to the problem of standardizing the kind of data stored according to principles used in the central databank for the Italian flora. At a meeting of interested users of the local banks in Rome in February 1982, it was agreed that nomenclature, coding of species, coding of environment types, etc., will follow the standards used in the central databank. Furthermore, the software adopted by the various local groups is basically the same, so that the coordination of the different local databanks within a common structure does not present the kind of problems that are usually met when different centres use different software and different coding procedures.

The basic plan is to connect the local databanks in the form of a distributed database. In this way, information stored independently by the different local units could be utilized by any other unit. The central databank for the Italian flora will maintain its dual role as a "safe" and as the main reference system for the standardization of parameters used in the local databanks. A series of interface programs are under development, connecting the network to the central bank, to compare the information inductively acquired and stored in the latter with information crystallized in the former. This will allow us possibly to change data in the central databank according to new information gained through the activity of local groups.

Finally, there is one point that we would like to stress: it is often assumed that databanking is a boring activity to be involved in, whose only practical result is to substitute traditional methods of data storage with more modern and efficient ones. As far as we can judge from the experience gained in the last 10 years, the introduction of databanks in the activity of taxonomists and vegetation scientists is likely to have a more profound impact. What will change is not only a certain style of work (the coordination of single researchers and groups will be more and more stressed in the future), but also the type of problems that will arise, whose importance is chiefly of methodological-epistemological character. One example, from phytosociology, is the evaluation of syntaxonomical schemes obtained through the numerical elaboration of large sets of relevés, another example is the shift toward quantification of data in phytogeographical studies. In conclusion, it is our opinion that databanks will not only prove to be time-saving instruments, but will also pay high "dividends" in terms of the definition and solution of purely scientific problems.

REFERENCES

Anzaldi, C. and Mirri, L. (1979). Un esperimento di strutturazione di dati floristici e vegetazionali. *Pubbl. IAC.* 195, 1–148.

Anzaldi, C. and Mirri, L. (1980). La banca dei dati floristici e vegetazionali: controllo e formalizzazione die dati di input. *Coll. Progr. Fin. Qualità Ambiente AQ/5/29*. Rome.
Brenan, J. P. M., Franks, J. W., Rajnal, J. and Cullen, J. (1975). Report of working party on electronic data processing in major European plant collections. *Adansonia* 15, 7–24.
Cristofolini, G., Lausi, D. and Pignatti, S. (1969). "Survey of the system for coding plant sociological records used by the Trieste-group". Trieste.
Ellenberg, H. and Cristofolini, G. (1964). Sichtlochkarten als. Hilfsmittel zur Ordnung und Auswertung von Vegetationsaufnahmen. *Ber. geobot. Forsch-Inst. Rübel* 35, 124–134.
Estabrook, G. F. and Brill, R. C. (1969). The theory of the Taxir accessioner. *Mathematical Biosc.* 5, 327–340.
Feoli, E. (1977). A criterion for monothetic classification of phytosociological entities based on species ordination. *Vegetatio* 33, 147–152.
Feoli, E. and Lagonegro, M. (1979). Intersection analysis in phytosociology: computer program and applications. *Vegetatio* 40, 55–59.
Feoli Chiapella, L. and Feoli, E. (1983). Predizione ambientale basata su flore locali. Un esempio di applicazione della banca dati TAXIR. *In* "Le comunità vegetali come indicatori ambientali" (C. Ferrari *et al.*, eds), pp. 111–131. Bologna.
Kortekaas, W. M., Lausi, D., Beetfink, W. G. and van der Maarel, E. (1976). Survey of salt marsh relevés included in the data bank of the working group for data-processing. *Vegetatio* 33, 51–60.
Lagonegro, M. and Feoli, E. (1979). Iahopa, a useful program for systematics and ecology. *Quad. CDC. Trieste* 12, 1–51.
Lagonegro, M., Ganis, P., Feoli, E., Poldini, L. and Canavese, T. (1982). Un software per banche dati di flore territoriali estendibile alla vegetazione. *Coll. Progr. Fin. Qualità Ambiente AQ/5/38*. Udine.
Lausi, D. and Feoli, E. (1979). Hierarchical classification of European salt marsh vegetation based on numerical methods. *Vegetatio* 39, 171–184.
Nimis, P. L. (1981a). La banca dati relativi alla flora e vegetazione d'Italia. *Quad. CNR* AC/1/105, 83–86.
Nimis, P. L. (1981b). La banca dati sulla flora e vegetazione d'Italia: utenza e gestione. *Quad. CNR* AC/5/14, 41–57.
Pignatti, S. (1973). Problemi di codifica dei dati floristici in fitosociologia. *Not. Fitosoc.* 7, 17–20.
Pignatti, S. (1975). Zum Stande der floristischen Kartierung Mitteleuropas in Norditalien. *Gött. Flor. Rundbr.* 2, 61–63.
Pignatti, S. (1976). A system for coding plant species for data-processing in phytosociology. *Vegetatio* 33, 23–32.
Pignatti, S. (1978). Dieci anni di cartografia floristica nell'Italia di Nord-Est. *Inf. Bot. Ital.* 10, 212–219.
Pignatti, S. (1980). Natural Vegetation. *In* "Italy, a Geographical Survey" [24th Georgr. Congress] (M. Pinna and D. Ruocco, eds), pp. 115–140. Pacini, Pisa.
Pignatti, S. (1981). Check-list of the flora of Italy with codified plant names for computer use. *Quad. CNR* AQ/5/13, 000.
Pignatti, S. (1982). "Flora d'Italia". Calderini, Bologna.
Pignatti, S. and Sauli, M. (1976). I tipi corologici della flora italiana e la loro

distribuzione regionale: elaborazione con computer di 2600 specie di Angiosperme dicotiledoni. *Arch. bot. biogeogr. ital.* 52, 117–134.
Pignatti, S., Cristofolini, G. and Lausi, D. (1968). Verwendungsmöglichkeiten einer elektronischen Datenverarbeitungsanlage für die pflanzensoziologischen Dokumentation. "Ber. Int. symp. f. Vegetationsk.", Rinteln.

11 Fact Documentation and Literature Database for the Crustacean Order Tanaidacea

J. SIEG

Universität Osnabrück/Abteilung Vechta, Fachbereich, Naturwissenschaften, Mathematik, Driverstraße 22, D-2848 Vechta, Germany

Abstract: Consideration of the special database requirements that taxonomists have shows that bibliographic databases are unsatisfactory. I therefore separate bibliographic information into "citations" and "facts", each component being stored in a different database: "citations" in the CRUSTACEA database and "facts" in the TANAIDACEA database. Since the CRUSTACEA database is a conventional bibliographic database, only a short description is given. The TANAIDACEA database is discussed more fully. The datasheet used, the data-management facilities incorporated and the requirements for database construction are presented. Common taxonomic problems (e.g. synonymy, wrong determination, etc.) cause great difficulty in linking documents with the correct taxa. Normally taxonomic codes are assigned numerically in hierarchical order. Because these codes are inflexible and inadequate, a combined alphabetic and numeric code is here proposed. Finally, some of the services available from the TANAIDACEA-database (e.g. distribution maps, bibliographic, synonymised and alphabetical lists of species) are illustrated and services planned for the future (e.g. museum catalogues and check-lists) are discussed briefly.

INTRODUCTION

The use of computer databases in taxonomy is not as common as in other sciences, their main usage having been in numerical taxonomy, the generation of keys, in biogeographic systems and in the curation

Systematics Association Special Volume No. 26, "Databases in Systematics", edited by R. Allkin and F. A. Bisby, 1984, pp. 125–136. Academic Press, London and Orlando.

DATABASES IN SYSTEMATICS
ISBN 0 12 053040 6

of herbaria or zoological collections. Recently, taxonomists discovered databases as a tool for tasks formerly done by hand. As microcomputers become cheap it is not surprising that many small and specialized databases have been constructed. Normally, these only contain information of direct interest to its creator and only scientists working in the same field make use of them. Because of the central position of taxonomy, databases established in this domain have to include as much information as possible. One of the goals of "fact documentation" was that scientists of other branches of biology could also use the system, thus justifying the huge investment of money and labour.

OVERALL PLAN FOR THE DOCUMENTATION SYSTEM

The plans for fact documentation for the Crustacean order Tanaidacea grew gradually through continual discussion of the type of information services that colleagues would like. We arrived at a long list of services which could be separated easily into two groups. For example, the request "all known literature for the Tanaidacea" is easier to answer than "all recorded occurrences of *Tanais dulongii*". The latter requires more detailed analysis of the literature and a totally different type of output. This led to the development of separate databases; the CRUSTACEA database for literature citations and the TANAIDACEA database for all factual information found in specialist papers. Limited transfer of information via scratch files should be possible.

A second step involved technical considerations. Database management software was poorly developed by Telefunken but the retrieval system implemented was adequate. So we had to search for additional software designed for handling documents. The only package currently running on the TR 440 is "TUBIBMUE" which was developed at the Technische Universität München (Langendörfer, 1973; Braun and Langendörfer, 1976; Halfar and Langendörfer, 1976). This package includes programs for automatic keywording and database management, including editing and deletion of items or descriptors (for more details see Aschenbrenner *et al.*, 1977). Unfortunately, it was written exclusively for bibliographic databases so the second major decision was to store initially only the bibliographic data in a database. Factual information was arranged in a sequential file. We were sure that this was only a transitional stage since an extension of "TUBIBMUE" was planned to handle both types of data.

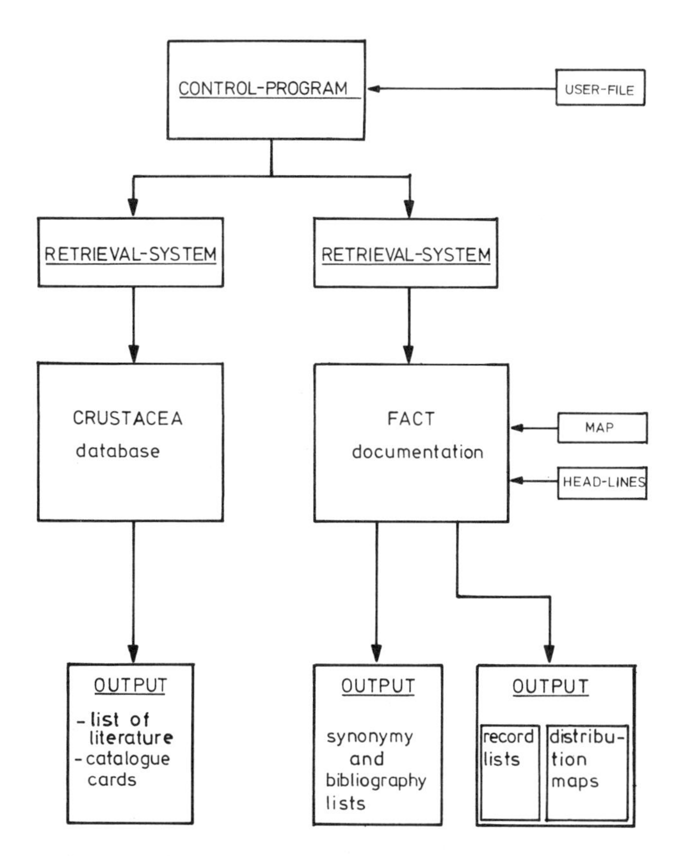

Fig. 1. General view of Tanaidacea documentation system.

PROCESSING OF LITERATURE

All papers referring to the Tanaidacea are processed but treated differently for the two databases: CRUSTACEA and TANAIDACEA.

Data capture for CRUSTACEA is, with minor exceptions, as described by Aschenbrenner *et al.* (1978). A modification which should be mentioned is "within keywords". Ordinarily keywords should be separated by commas, but we also allow one or more keywords in parentheses, e.g. Tanadacea (morphology, distribution),

Tanaiscavolinii, etc. During processing, keywords in front of parentheses are linked with those inside (e.g. Tanadaceamorphology). Linked keywords are both stored separately and linked together (e.g. Tanaidacea, Tanaidaceamorphology, morphology and Tanaidaceadistribution). Thus the retrieval system is able to distinguish between a paper dealing with the morphology of tanaids and one referring to decapod morphology.

For the TANAIDACEA database a comprehensive datasheet is completed for each taxon mentioned in a paper. The field "citation" contains author, year, journal or book, volume and pages. The category "remarks" represents an abstract of all information concerning the taxon in question (maximum size: 36 lines of 75 characters each). "Remarks" is followed by taxon information, collector, determinator, date (when the specimens were collected), location and status (museum), as well as further specifications such as number of specimens, larvae or adults. The field "location" contains a textual description with the geographical coordinates and geo-code. Because tanaids are a group of marine Crustaceans, we also have the fields "salinity" and "depth". Last, but not least, we have "additional remarks" for extra information, generally concerning the location (more detailed descriptions, sediment analyses, etc.). Data input for both databases is on-line to keep data errors to a minimum.

THE CRUSTACEA DATABASE

This is a normal bibliographic database and, therefore, only a short description will be given here (for more details see Aschenbrenner *et al.*, 1977). Data processing is as for other systems and has been described in detail by Braun and Langendörfer (1976). Following initial corrections to spelling mistakes, etc., using an EDITOR, the format and completeness of the data stored are checked. The software (DOKUM) includes automatic indexing, identification of language (English, French, German) and production of keyword lists from the title, keywords and free text. Meaningless words to be suppressed from all three languages are listed in alphabetical order. Finally, since information is still in a readable form, it can be corrected using an EDITOR before the data are entered into the database.

CRUSTACEA is arranged hierarchically and divided into the thesaurus and reference parts. The thesaurus is structured using different categories (e.g. author, year, etc.). In addition cross-references are necessary because the title of each paper is stored in its original language and keywords are not standardized. Cross-references include synonyms, antonyms, main and subordinate term, semantic field and homonyms (Aschenbrenner *et al.*, 1977).

The retrieval system at our disposal (TELDOK, see user manual: Computer Gesellschaft Konstanz, 1975) allows logical ("and", "or" and "not") and weighted ($b \neq 0, -127 \leqslant b \leqslant +127$) retrieval. Besides many other possibilities, selected articles can be transferred to a scratch-file which is then input to a printing program developed by us. Two different styles of output, "literature lists" and "catalogue cards", are currently possible.

Two developments are planned. The present format for the catalogue cards (A-6) will be supplemented by "international library catalogue cards". The "descriptor" program in preparation is of much more interest: it will add descriptor lists with cross-references (document-number) to either the literature list or the catalogue cards.

THE TANAIDACEA DATABASE

Planning the TANAIDACEA database was more involved because of its more complex data structure. Even when we referred to only one type of article, we could find no adequate software system. Data were therefore stored in a sequential file and searched using a retrieval-system written by ourselves. Our problems are now solved since an extended version of DOKUM is available and transfer of data is scheduled for 1983. Data processing will be analogous to that for the CRUSTACEA database.

During our discussions of the general requirements for a system to handle factual information it became obvious that much greater flexibility was required than available in any system known to us. The principal reason for creating databases previously has been to acquire knowledge of the distribution of species: the aim being to print maps or faunal lists. Bibliographic data was included only for cross-reference (Haeupler *et al.*, 1976; Heath, 1971; Jungbluth *et al.*, 1981). Believing that all available information should be accessible, it was decided to include abstract type information for each taxon mentioned in a paper to make the database attractive to scientists other than taxonomists or biogeographers. For example, it will also be possible to find out if certain structures (e.g. alimentary glands) have been described, illustrated or investigated using electron microscopy, etc.

Because of common taxonomic problems such as synonymy, mis-identification, etc., great difficulties exist in associating articles with the correct taxa. Normally for this purpose taxonomic codes are used, which number taxa in their hierarchical order. Because these codes are inflexible and inadequate, a combined system of alphabetical names and numerical codes is used. Taxa above species level are defined by the names of taxa included and species are defined using a sequence of numerical codes.

All documents referring to a certain species will have a nine-digit code number attached. The first four figures were obtained by arranging all known tanaid genera in alphabetical order and

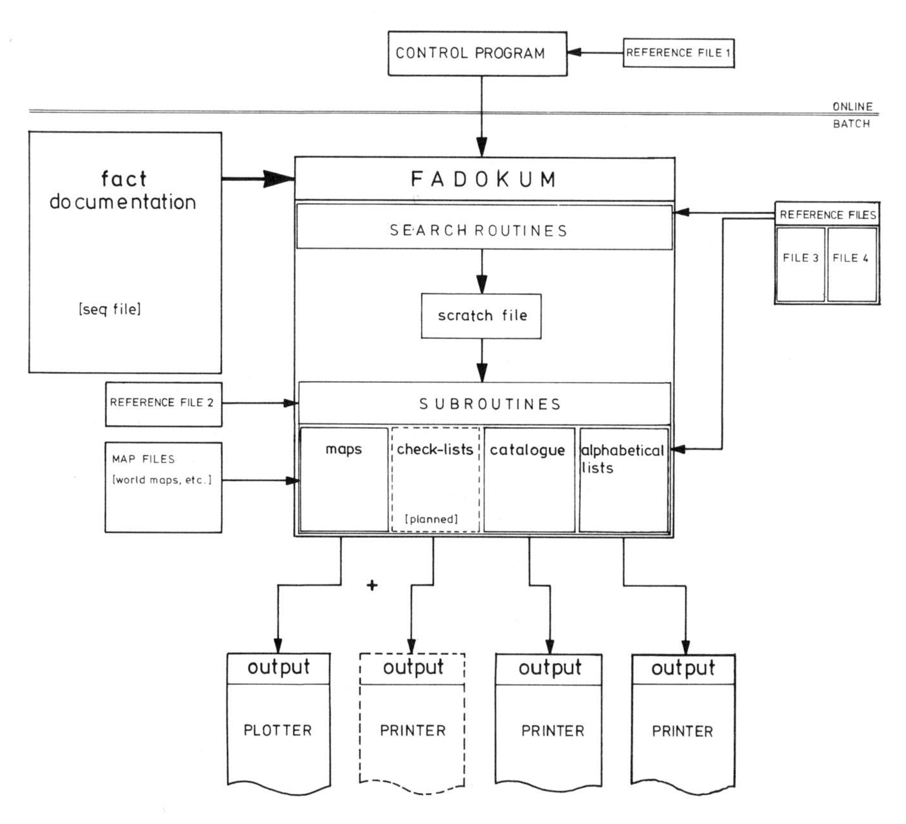

Fig. 2. "Fact documentation".

numbering them in steps of ten (001000000–899900000). All taxa above genus level start with 9 as the first digit (e.g. Apseudidae = 906000000). Species belonging to the same genus were also arranged in alphabetical order and numbered using the following three figures in the nine-digit code (e.g. *Apseudes* = 013000000, *Apseudes spinosus* = 013084000). The remaining two figures are used to indicate spelling variants (e.g. *Heterotanais örstedi* = 047014007; *Heterotanais oerstedi* = 047014043) or for different delimitations ("sensu"-variants) of a taxon. This arrangement allows great flexibility when printing.

During the last three years emphasis has been placed mainly on storing all the available data and, therefore, we were not able to make full use of available facilities. Three facilities have been established so far: catalogues, distribution maps and alphabetical lists (Fig. 2).

1. Catalogues

This subroutine produces output similar to the CATALOGUS CRUSTACEORUM (e.g. Sieg, 1983). As mentioned previously, taxa

1. Gattung: Anatanais NORDENSTAM, 1930

Tanais (partim). - GERSTAECKER, A., 1883-1888, IN BRONN: KLASSEN UND ORDNUNGEN DES THIER-REICHES 5 (2): 517 (ALLGEMEINE AMMERKUNGEN, ZUM TEIL VERSTREUT).

Tanais (partim). - NIERSTRASZ, H.F., 1913, SIBOGA EXPEDIT. 32 (A): 20-32 (AUFLISTUNG DER ARTEN). - SIEG, J., 1973, ZUM NATUERLICHEN SYSTEM DER DIKONOPHORA LANG -: 35, 77 (KURZE ANMERKUNGEN ZUM BAU DES PLEONS). - SIEG, J., 1976, Z. ZOOL. SYST. EVOLUT.-FORSCH. 14: 179 (KURZE ANMERKUNGEN ZUM BAU DES PLEONS).

Anatanais subgen. NORDENSTRAM, A., 1930, NAT. HIST. JUAN FERNANDEZ EASTER ISL., (ZOOL.) 3: 525 (KURZDIAGNOSE; AUFLISTUNG DER ZUGEHOERIGEN ARTEN, TYPUS-ART WIRD JEDOCH NICHT ANGEGEBEN). - MILLER, M.A., 1940, OCC. PAP. B. P. BISHOP MUS., 15: 301 (KURZE ANMERKUNGEN). - LARWOOD, H.J.C., 1954, ANN. MAG. NAT. HIST., (12) 7: 569 (LITERATURZITATE; KURZE ANMERKUNGEN; UMFANG).

Typus-Art: Tanais (Anatanais) lineatus NORDENSTAM, 1930
(durch spaetere Festlegung; SIEG, 1980, im Druck).
Geschlecht: Maennlich.

Fig. 3. Normal output from the catalogue subroutine.

above the species level are defined by the species included and the sequence of species names defines their arrangement upon output. Thus taxa can be listed in systematic order (e.g. Fig. 3). Within each taxon, citations are printed in chronological order, articles published in the same year being arranged alphabetically by taxon name. The headings used in the printed output (Fig. 3) are stored in a separate file and are linked using the same code numbers. A second version of the catalogue subroutine, called "chronological", arranges all articles including a given taxon by year of publication. Within the same year, names of taxa are sorted alphabetically (Fig. 4).

2. Art: Anatanais novaezealandiae (THOMSON, 1879).

1879 Tanais novae-zealandiae THOMSON, G. M., ANN. MAG. NAT. HIST., (5) 4: 417 (KURZE BESCHREIBUNG; ABBILDUNGEN AUF PLATE XIX FIG. 5 LATERALANSICHT, FIG. 6: P. 6; OB DAS TYPEN-MATERIAL NOCH VORHANDEN IST, KONTE BIS JETZT NICHT ERMITTELT WERDEN; ALS LOCUS TYPICUS IST DUNEDIN HARBOUR - 8 BIS 10 M - ANGEGEBEN).
1880 Tanais novae-zealandiae THOMSON, G. M., ANN. MAG. NAT. HIST., (5) 6: 207 (WIEDERHOLUNG DER ANGABEN VON 1879: 417; ABBILDUNGEN AUF PLATE VII FIG. 3: LATERALANSICHT).
1882 Tanais novaezealandiae.- SARS, G. O., ARCH. MATH. NATURV., 7: 24 (LITERATURZITAT; VERBREITUNGSDATEN).
1886 Tanais novaezealandiae.- SARS, G. O., ARCH. MATH. NATURV., 11: 311 (VERBLEIBT TROTZ STARKER EINENGUNG DER GATTUNG TANAIS IN DIESER).
1887 Tanais neo-zelanica.- THOMSON, G. M. & CHILTON, C., TRANS. PROC. N. Z. INST., 18: 151 (LITERATURZITATE; VERBREITUNGSDATEN, NEUER FUNDORT: LYTTELTON; ANM.: UNGERECHTFERTIGTE EMENDATION, ALSO JUENGERES OBJEKTIVES SYNONYM).
1896 Tanais novae-zealandiae SARS, G. O., AN ACCOUNT OF THE CRUSTACEA OF NORWAY, 2 (ISOPODA): 12 (DIE ART VERBLEIBT IN DER GATTUNG TANAIS).

Fig. 4. "Chronological" output from the catalogue subroutine.

The amount of information printed can be varied. The most comprehensive version prints author, year, citation (including pages, but not the title of the paper) and remarks. This can be reduced so that only author, year and citation or just author and year appear.

2. Distribution Maps

The UTM-grids used in the European Invertebrate Survey, were not suitable for the Tanaidacea, which have a worldwide distribution. Since tanaids are a marine group, their location is nearly always recorded using a ship's position (normally specified by geographical coordinates). Knowing the difficulties in transforming these data to UTM-grids, we decided to employ the original coordinates.

Fortunately, it was not necessary to digitize a world map since this was available from the Institut für Angewandte Geodäsie, Frankfurt. An advantage of storing data as geographical coordinates is that special types of map can easily be obtained if their projection formula is known. For the TANAIDACEA database four different maps have been produced: two world maps (HAMMER-projection, zylinder projection) and two maps of the poles (Arctis, Antarctis). Many different maps can be produced by varying the two fields "taxon" and "geo-code/location".

The zoogeographical areas proposed by DeLattin (1967) for the world oceans are assigned code numbers in a hierarchical fashion. This makes it possible to plot maps for the tropical warm water area (e.g. geo-code 3000) as well as only for the Atlantic section (e.g. 3400) or only for the shallow waters of the west-coast of North Africa (e.g. 3410). Similarly maps can be defined using the field "taxon". For taxa above the species level there is a choice between "all records in one map" or "separate maps for each species included". In this way analysis of distribution patterns can be done both for a family and for individual species. Doubtful records are indicated by question marks. "Taxon" and "geo-code" are interlinked so that the program allows not only separate maps for each subregion, but also separate maps for each species within each subregion.

3. Alphabetical Lists

The literature of tanaids is widely scattered and it is difficult to obtain a complete list of those species known to science. Unfortunately their taxonomy is also unstable. It is not uncommon for a species described in a certain genus to be transferred some months later to a totally different genus. To keep scientists informed of these changes, we wrote a subroutine to produce alphabetical lists for any desired taxon (Fig. 5), giving the code numbers and synonymies for all valid species.

CODE-NR	TAXON	GEO-CODE
0020000	Agathotanais HANSEN, 1913.	
0020010	Agathotanais hanseni LANG, 1971.	0000, 0000, 0000
0020020	Agathotanais ingolfi HANSEN, 1913.	0000, 0000, 0000
0020030	Agathotanais splendidus KUDINOVA-PASTERNAK, 1968.	0000, 0000, 0000
0030000	Akanthinotanais SIEG, 1973.	
0040000	Alaotanais (siehe: Neotanais BEDDARD, 1886.)	
0045000	Allotanais SHIINO, 1979. (syn.: Archaeotanais)	
0045010	Allotanais hirsutus (BEDDARD, 1886). (syn.: Anatanais hirsutus Anatanais nierstraszi Archaeotanais hirsutus Tanais hirsutus Tanais nierstraszi)	0000,0000, 0000
0050000	Anarthrura G. O. SARS, 1882.	
0050020	Anarthrura simplex G. O. SARS, 1882.	0000 0000, 0000

Fig. 5. Portion of the alphabetical list produced by "fact documentation".

FUTURE PLANS FOR "FACT DOCUMENTATION"

After five years' work, phase II of the project (data collection) is finished. We have now started to use the stored information (phase III).

No further development of the catalogue routine is planned but many changes are necessary to the mapping routine. At present it is impossible to print distribution maps in which records for different species can be distinguished. Using a characteristic symbol for each species, an extended version will allow up to nine species to be included in one map. Thereby the basis will be laid for plotting maps for particular depths or times. It will be possible to choose between separate maps for each period/depth-range and one map in which the different periods/depth-ranges are indicated by special symbols. It further seems reasonable to produce maps for different geographical areas to replace the world map if necessary. Finally we are to write a subroutine to transcribe distribution data into a format for use by other programs such as SYMAP, ASPEX or GENSTAT.

Another urgent need is for distribution maps with written information. An additional subroutine is planned which will print checklists and faunistic lists. Finally a more comprehensive subroutine is planned to use the field "museum" to print catalogue cards or "type-material" catalogues for those specimens or species held by a given museum.

ACKNOWLEDGEMENTS

Sincere thanks are due to Dipl.-Math. W. Bauch, who did most of the programming for the fact documentation, and to W. Hanken, who wrote the printing routine for the CRUSTACEA database. Dr I. Pfrommer and Dipl.-Ing. I. Wilsky (Institut für Angewandte Geodäsie/Frankfurt) kindly made the database "Welkarte" available. I am deeply indebted to Prof. H. Langendörfer (München) who mainly developed the package "TUBIBMUE" and to Dipl.-Math. T. Haarman and Dipl.-Math. A. Schütt of the Rechenzentrum der Universität Osnabrück for implementing it. This work was supported by grants from the German government.

REFERENCES

Aschenbrenner, M., Hein, M. and Mauer, R. (1977). "Das Bibliotheks-Dokumentations-System TUBIBMUE - Aufbau und Wartung -." Institute für Informatik, Technische Universität München [Bericht 7704].

Aschenbrenner, M. *et al.* (1978). Datenerfassung für ein Bibliotheks-Dokumentations-system. Institut für Informatik, Technische Universität München [Bericht 7819].

Braun, S. and Langendörfer, H. (1976). Aufbau eines Dokumentations-systems für eine Fachbibliothek. *In* "Praxis der Realisierung von Informationssystemen" (F. Haßeld, ed.). Hanser Verlag.

Computer Gesellschaft Konstanz. (1975). "TELDOK-Benutzerbeschreibung, April 1975".

DeLattin, G. (1967). "Grundriss der Zoogeographie". Fischer, Jena.

Haeupler, H., Fink, H., Hamann, U., Hansen, K., Krach, E., Niklfeld, H., Schönfelder, P. and Schroeder, F.-G. (1976). "Grundlagen und Arbeitsmethoden für die Kartierung der Flora Mitteleuropas (2. erweiterte Auflage)". Zentralstelle für die floristische Kartierung Westdeutschland, Göttingen.

Halfar, H. and Langendörfer, H. (1976). "Benutzeranleitung zum Recherchieren in der Datenbank TUBIBMUE". Institut für Informatik, Technische Universität München [Bericht 7620].

Heath, J. (1971). "European Invertebrate Survey. Instructions for recorders". Biological Records Centre, Monks Wood Experimental Station, Abbots Ripton, Huntingdonshire.

Jungbluth, J. H., Bürk, R. and Berger, J. (1981). Zehn Jahre Molluskenkartierung in der Bundesrepublik Deutschland - Beispiel einer faunistischen Modellkartierung. *Natur und Landschaft* 57, 309–318.

Langendörfer, H. (1973). "Eine Syntax zur Beschreibung des Literaturbestandes einer Bibliothek". Abteilung Mathematik Technische Universität München [Bericht 7308].

Sieg, J. (1983). Tanaidacea. *In* "Catalogus Crustaceorum" (H. E. Gruner and L. B. Holthuis, eds), part 6, 1–552. The Hague, Netherlands.

12 PRECIS — A Curatorial and Biogeographic System

G. E. GIBBS RUSSELL and P. GONSALVES

Botanical Research Institute, Private Bag X101, Pretoria, Republic of South Africa

Abstract: PRECIS (Pretoria Computerized Information System) is a databank of herbarium specimen label information. All the specimens of indigenous and naturalized plants at the National Herbarium, Pretoria, from the territories covered by the Flora of Southern Africa, are included in the system. To date there are about 620 000 specimens in PRECIS, and about 20 000 accessions are added annually. The system is maintained on the Burroughs 7800 computer of the Department of Agriculture. The specimens are arranged in ten datasets of roughly equivalent size that correspond to the order of specimens in the Herbarium. There are additional datasets for collectors, taxon names, computer specimen numbers, new specimens and unidentified specimens. The data for each specimen is stored in 24 fields, which record the collector and number, place and date of collection, height, brief notes about the habitat, current identification, type status, state of flowering and fruiting, and herbarium code. Some of the fields are numeric and others are alphanumeric; some fields are coded and others are free format. Output from the system is of two kinds: (1) by means of Burroughs language DMS-INQUIRY, through which specimens may be selected and sorted on any field, and any desired fields reported; and (2) by means of special programs written for the system. These special programs provide specimen labels as well as listings of specimens in taxa, names of taxa, specimens or species in latitude and longitude grids, collectors and localities.

HISTORY OF PRECIS

During the early 1970s plans for a herbarium databank at the National Herbarium at the Botanical Research Institute in Pretoria

Systematics Association Special Volume No. 26, "Databases in Systematics", edited by R. Allkin and F. A. Bisby, 1984, pp. 137–153. Academic Press, London and Orlando.

DATABASES IN SYSTEMATICS
ISBN 0 12 053040 6

(PRE) were begun. The first step was to provide a simple grid reference system (Edwards and Leistner, 1971), and gazetteer of place names (Leistner and Morris, 1976), so that localities could be recalled easily by the computer for the preparation of distribution maps and check lists (Morris and Leistner, 1971).

At the same time, plans for the content and format for a herbarium computer system were made, on the basis of development in herbarium systems elsewhere and the computer facilities available (at that time a Burroughs 6700 computer) (Morris, 1973, 1974). Encoding began in June 1975, and by December 1976 the *c*. 470 000 backlog specimens had been processed. The work was completed by 28 persons working 12–16 hours per week in the evenings. The encoders were not botanically trained, but the supervisors were botanists (Morris and Leistner, 1975a, 1975b; Morris, 1980).

Morris and Glen (1978) described the PRECIS system as it was originally envisaged. It was to have encompassed three bases, each composed of several datasets:

(1) literature base, whose most important file was literature citations;
(2) taxon base, whose most important file was genus and species, and which formed the link between all three bases;
(3) specimen base, whose most important file was specimens.

The complex specimen base with about 50 fields was developed first.

Development of the programs took place concurrently with specimen encoding and in 1979 the system was handed over to the Botanical Research Institute as operational. By this time, collectors were routinely using encoding forms for new collections, and three assistants encoded label data for specimens from outside the Institute. Output could be obtained in two formats:

(1) "minis", with five lines of data about the specimen, consisting of collector, identification and locality information; or
(2) "maxis", which reported all the data stored for the specimen.

The detailed report on the contents of PRECIS by Morris and Manders (1981) also showed potential uses of the system. Their analysis of specimen data indicated that only about 10% of the specimens had grid references, although this information had been seen from the beginning as crucial for the applications (Morris and Leistner, 1971). Consequently, two major operations were undertaken to add grid references. The grid for every locality that was recorded on more than 20 specimens was found, and added to all the specimens from these localities.

In the second addition of grids the *c*. 50 000 grass specimens were listed by locality. Herbarium professional staff checked the grids that were already present, and looked up grids for specimens where none was recorded. Localities which could be accurately pinpointed were added for all specimens and localities which could be determined only for an individual specimen were added to that specimen alone. At the end of this about 80% of all grass specimens and about 65% of all others had a grid reference in PRECIS.

During the first year of production, it became evident that difficulties with the maintenance and operation of PRECIS reduced its value to the Institute. These problems were defined by Magill *et al*. (in press):

(1) The system was too large. It occupied four disc packs on the Burroughs computer, so that it could be operated only at night. Morris and Manders (1981) showed that this was due in part to empty records and in part to extraneous and repetitive information that resulted from mixing specimen and taxon data.
(2) The database was too complex. Program maintenance and alterations were difficult, and information retrieval was lengthy and expensive because the data for every specimen were split into a number of different datasets.
(3) The output format was too inflexible. For many queries, the entire amount of data reported on the "minis" or "maxis" was not required, but these were produced nevertheless, thus increasing the cost of the inquiry.
(4) The specimen labels were in cryptic format. Much of the data was recorded in PRECIS as codes, which were retranslated into words for printing on specimen labels. This resulted in the label information being printed in a disjointed and ungrammatical form that was not suitable for the herbarium.
(5) Users were discouraged from using the system. The high cost of routine inquiries, the practical difficulties of loading, manipulating and correcting the data, and the slow turnaround of requests caused most of the Institute's curatorial staff and researchers to be disillusioned with the system.

In the light of these difficulties, the Institute reassessed the system with special emphasis on the needs of herbarium curation and flora research. Datametrical Services of the Department of Agriculture made suggestions for its improved operation and efficiency. As a result, the database was completely restructured during nine months by an intensive cooperative effort between the Institute and the Datametrical Services and Computer sections of the Department of Agriculture. Ecological, economic, photographic, horticultural and bibliographic information was discarded from PRECIS, and only specimen-based data useful for herbarium curation and taxonomic research were retained.

Table 1. Comparison between the original system ("PRECIS I") and the present system ("PRECIS II"). R = Rand.

	PRECIS I	*PRECIS II*
Number of disc packs	4	1
Mode of operation	batch	batch and on line
Mode of storing data for each specimen	Each specimen fragmented into a number of datasets	Each specimen entire, in 1 of 10 taxonomic datasets
Format of output	"Maxis" and "Minis" only	"Minis" Special listings INQUIRY reports
Cost of listing species in all 1/4 degree squares	(operation not done in PRECIS I)	R 600
Cost of listing all specimens in a dataset in taxonomic order (example: Poaceae, 50 000 specimens)	R 4 200	R 210
Cost of security back-up per month	R 8 000	R 200

Table 1 shows a comparison between the original system ("PRECIS I") and the system which is now in use ("PRECIS II"). The new system, which has been in operation since April 1982 is described below.

DESCRIPTION OF PRECIS II

The core of PRECIS II is ten large specimen datasets, with two smaller datasets for new and for unidentified specimens. Peripheral to this are the taxonomic and specimen-related datasets for plant collectors, current species names and registered collecting localities. In terms of the operation of the system, the collecting localities, new specimens and unidentified specimens datasets are temporary, whereas the rest are permanent. For each dataset there is an updating program and one or more listing programs.

Data fields for each specimen are encoded and stored primarily as alphanumeric data. In PRECIS I, specimen fields were mostly numeric codes, and the corresponding alphanumeric data for each code was stored in a separate dataset.

1. Specimen Encoding Form

The specimen encoding form is divided into four sections. Section A identifies the specimen by collector, number, data and locality. Section B records notes about the habitat and appearance of the plant. Section C contains codes for storing habitat data in the data bank. Section D records curatorial information. In this section the most important fields are the GENSPEC, which is the code used to retrieve the full name attached to the specimen, and the PRENO, which is a unique number used to locate the specimen in the system.

The form is A5 in size, so that it is convenient to use in the field. To further aid collectors, instructions for using the form are printed on the back.

2. Details of Datasets

All the data on the encoding form is initially loaded into the temporary specimen dataset NEWSPEC, and kept there until the labels have been printed. When the labels have been printed, the specimens are transferred to their permanent datasets, and some of the data on the encoding form are discarded. The notes in Section B and determinavit, date and number of labels needed from Section D appear on the specimen labels, but are not permanently stored in the system.

The ten permanent specimen datasets follow the Englerian system used in the Herbarium, and each of the wings in the herbarium building has two or three datasets. Most of the datasets include a number of plant families, and the name of each dataset is taken from the names of the first and last families in it. However, GRASSE, LEGUME and COMPOS each have only one large family.

The temporary specimen dataset UNIDENT holds specimens that did not have valid names when they were transferred from PRECIS I to PRECIS II. These will be gradually worked into their correct permanent dataset.

Other permanent datasets hold specimen-related information. There are four GENSPEC datasets that each contain the full scientific names, with authorities, for all the species in one of the four herbarium wings. With each name is also stored the code (GENSPEC number) for that name, and the number of specimens in the system assigned to the name. The maximum space allowed for each name is 120 characters. A single COLLECTOR dataset records the name, with initials, of all the collectors in the system, with a maximum of 60 characters allowed for each. The code for each collector is known as the collector's registered number (COLLREGNO), and is a five-digit number. The number of specimens recorded for each collector is also stored in the dataset.

REGLOCS is a temporary dataset of registered collecting localities for use in printing specimen labels. Its role is discussed in more detail below.

3. Loading Specimen Data and Printing Herbarium Labels

A batch system is used to load new specimens into the databank, and is summarized in Fig. 1. The information is encoded, punched and written into a data file. The program SPECUPDATE loads the specimens into the dataset NEWSPEC. "Counterfeit" herbarium labels on plain paper are printed for all specimens successfully loaded. If mistakes are found in proof-reading the "counterfeits" (Fig. 2) the specimens in NEWSPEC are updated. When all is correct, permanent ("goldplated") herbarium labels are produced on pre-printed label blanks (Fig. 3) and the specimens transferred from NEWSPEC into the permanent specimen datasets.

The REGLOC dataset contains all the data in Section A of the encoding form, except specimen number. When a collector takes more than one specimen at a particular site, one encoding form is used to record the details of the locality in Section A. This data is loaded into REGLOC. Other data for each specimen is encoded on subsequent encoding forms, but the locality data is not repeated on each of these specimen forms. The field "Registered Locality

Number" serves to tie each specimen encoding form to its locality data. When a specimen is loaded into NEWSPEC, the full locality data is obtained from REGLOC and stored with the specimen.

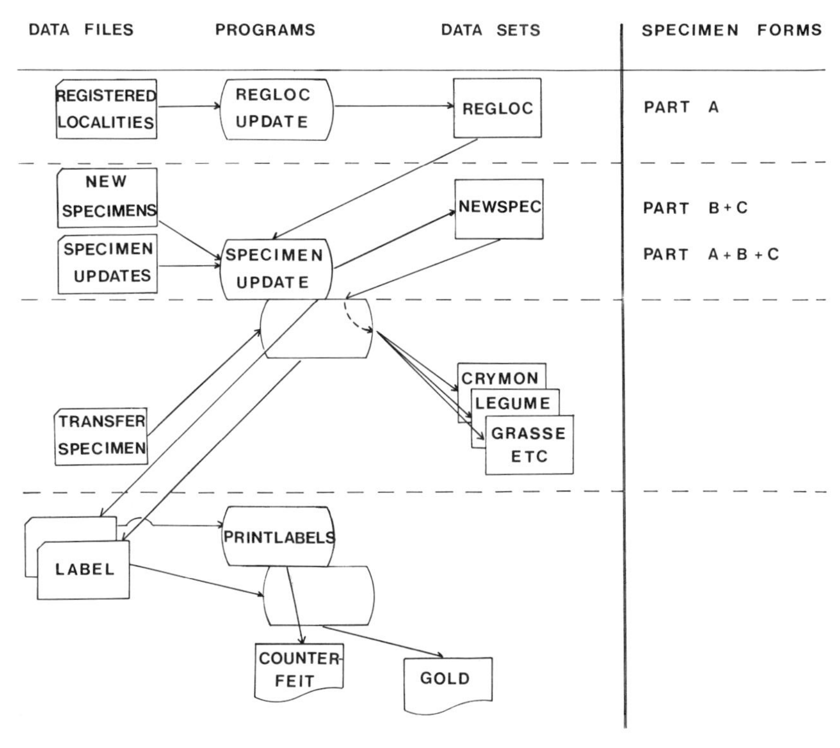

Fig. 1. Outline of data files, programs and datasets involved in loading new specimens and printing herbarium labels.

4. *Extracting Data*

Data is extracted from PRECIS with standard listing programs and with Burroughs DMS-INQUIRY system. A standard listing program lists the data in a particular dataset in a fixed format. For the specimen datasets, a listing of all the data for each of the specimens is referred to as a "mini" (a name carried over from PRECIS I). "Minis" may be printed in sequence by genus and species, by collector and specimen number, or by PRENO. Also for the specimen datasets, the SPECIES PER GRID program produces a list of species recorded for

0600862
2530DB TRANS
KRUGER PR
1979/06/06
359

TRICALYSIA CAPENSIS (MEISN.) SIM

NELSPRUIT.
ON NELSPRUIT - KAAPSEHOOP ROAD. 2 KM
ABOVE STELLA MINE.
S. MOUNTAIN SLOPE. ESCARPMENT. ROCKY
MOIST AREA. DARK BROWN LOAM. CLUMPS
OF DENSE BUSH WITH OPEN AREAS IN
BETWEEN. TREE. AMONGST LARGE GRANITE
BOULDERS IN AND AROUND CLUMPS OF
BUSH. FREQUENT 4-(6M) HIGH.
HEIGHT 4.00 M
P 00 OP SH STONY LOAM TREE
M. WELMAN. 0000 8308.000-00500

0600863
3322DC CAPE
VENTER HJT
1973/04/10
761

GALOPINA CIRCAEOIDES
THUNB.

WILDERNESS.

WOUDRAND. KRUIDAGTIG.

P 0 0 FORES HERB
M. WELMAN. 0000 8435.000-00200

BATCHNO: 0000045
ACCDATE: 07/82
LABELS: 01
LOCNO: 790000

BATCHNO: 0000045
ACCDATE: 07/82
LABELS: 01
LOCNO: 730000

Fig. 2. "Counterfeit" herbarium specimen labels printed on plain paper for proof-reading. Six labels are printed on a page, with details of batch, etc. in the right-hand margin. The penultimate line is a transcription of the habitat codes, which appear on the "counterfeit" for checking, but are not printed on the final ("goldplated") copy.

NATIONAL NASIONALE HERBARIUM PRETORIA 0601634

3029DA Grid Ref/Ruitverw. Regio NATAL
ARNOLD TH
0000001331 Legit & No. Anno Alt. 1980/01/18

PAPILLARIA AFRICANA (C. MUELL.) JAEG.

INGLI FOR.
4 KM FROM WEZA. BETWEEN KOKSTAD AND WEZA.

GROWING ON TREE TRUNK.
IN SHADED FOREST.

J. V. ROOY 1980 Det. Ref/Verw. 0001.743-00100

NATIONAL NASIONALE HERBARIUM PRETORIA 0601636

3029DA Grid Ref/Ruitverw. Regio NATAL
ARNOLD TH
0000001329 Legit & No. Anno Alt. 1980/01/18

FISSIDENS GLAUCESCENS HORNSCH.

INGELI FOREST.
BETWEEN KOKSTAD AND WEZA. C. 14 KM FROM WEZA.
DAMP BANK ON SIDE OF ROAD THROUGH FOREST.
SHADED.

J. V. ROOY. 1981 Det. Ref/Verw. 0001.316-01900

Fig. 3. "Goldplated" herbarium specimen labels, for attaching to the plant specimen and for distribution with duplicate specimens. Note that the computer specimen number (PRENO) appears at the top right, and the genus and species code (GENSPEC no.) appears at the bottom right.

each quarter degree square grid, as well as the number of specimens recorded for each grid. This program may be run on any of the permanent specimen datasets individually, or it may be run globally on all the datasets combined.

The standard listings for the dataset COLLECTORS gives the complete name of the collector, with the registered number and the number of specimens in the system for that collector. A listing may also be obtained of all collectors in numerical order by registered number.

The standard listing for the GENSPEC datasets yields a list of all plant species names in the herbarium wings. A program presents the list either in taxonomic order by GENSPEC number or in alphabetical order by genus. The number of specimens assigned to each name is given on both listings.

The Burroughs DMS-INQUIRY system offers a flexible procedure for extracting information from the databank. Specimens may be selected and sorted on any field, and fields may be selected individually or in combination. The selected specimens may be sorted in any designated order. Furthermore, any fields may be reported, the only restriction being that a single line of data may be displayed for a specimen. Examples of INQUIRY reports are shown in Figs. 4 and 5. Figure 4 shows a collector's register, giving all the specimens in the system for a particular collector, in sequence by specimen number. Figure 5 shows a list of specimens in the family Compositae from the high altitudes of the Natal Drakensberg, arranged in altitude groups.

5. Quality Control

The integrity of the databank is checked by the program AUTO-GLOBAL, but the accuracy of the data in a specific field is checked manually. A count of the number of items in each dataset is stored as global data and all programs that modify a dataset will affect at least one of these counts. If the counts of the program do not match the counts stored in the global data, then a warning message is printed. For example, the genus *Strychnos* belongs to the dataset OCHASC, and *Canthium* to BIGOOD. An inexperienced taxonomist may confuse the two genera in the sterile state. A *Canthium* specimen that was originally misidentified as a *Strychnos* must be transferred from one dataset to another, and the count of the number of specimens in each dataset will be affected.

COLLECTOR'S REGISTER FOR RE MAGILL

11/09/82

SPECIMEN-NO	NAME	YR	MO	DY	REGION	GRID	MAJOR	MINOR
5912	BRYUM SP.	1979	1	16	CAPE	3124AA		MOORDENAARSKLOOF
5915	DESMATODON CONVOLUTUS (BRID.) GROUT	1979	1	17	CAPE	3323AB	WILLOWMORE	AASVOELBERG
5916	TORTULA RURALIS (HEDW.) GAERTN., MEYER & S	1979	1	17	CAPE	3323AB	WILLOWMORE	AASVOELBERG
5917	HYPODONTIUM DREGEI (HORNSCH.) C. MUELL.	1979	1	17	CAPE	3323AB	AASVOELBERG	WILLOWMORE
5918	BRYUM CANARIENSE BRID.	1979	1	17	CAPE	3323AB	AASVOELBERG	WILLOWMORE
5919	HYPODONTIUM DREGEI (HORNSCH.) C. MUELL.	1979	1	17	CAPE	3323AB	WILLOWMORE	AASVOELBERG
5920	BARBULA SP.	1979	1	17	CAPE	3323AB	AASVOELBERG	WILLOWMORE
5921	GRIMMIA LAEVIGATA (BRID.) BRID.	1979	1	17	CAPE	3323AB	WILLOWMORE	AASVOELBERG
5926	TRICHOSTOMUM BRACHYDONTIUM BRUCH.	1979	1	18	CAPE	3323CA		UNIONDALE
5927	TORTULA PILIFERA HOOK	1979	1	18	CAPE	3323CA	LITTLE KAROO	UNIONDALE
5927	TORTULA PILIFERA HOOK	1979	1	18	CAPE	3323CA	LITTLE KAROO	UNIONDALE
5927	BARBULA SP.	1979	1	18	CAPE	3323CA	LITTLE KAROO	UNIONDALE
5940	BRACHYMENIUM PULCHRUM HOOK.	1979	1	18	CAPE	3323CC	UNIONDALE	PRINCE ALFREDS PASS
5942	CAMPYLOPUS STENOPELMA (C. MUELL.) REHM. EX	1979	1	18	CAPE	3323CC	UNIONDALE	PRINCE ALFRED PASS

Fig. 4. INQUIRY output. The query was structured in the following way. Dataset: CRYMON. Select: Collector = R. E. Magill. Sort: by specimen number. Report: Specimen number, scientific name, date, region, grid, major locality, minor locality.

COMPOSITAE OF THE HIGH DRAKENSBERG

11/09/82

NAME	SUBSTRATE	LIFEFORM	ALTITUDE	ASPECT
HELICHRYSUM PSILOLEPIS HARV.			2500	
HELICHRYSUM SP.			2500	
HELICHRYSUM SUBGLOMERATUM LESS. VAR. SUBGL			2500	
HELICHRYSUM SUTHERLANDII HARV.			2500	
HELICHRYSUM SUTHERLANDII HARV.			2500	
HELICHRYSUM TENUICULUM DC.	STONY		2500	
HELICHRYSUM TRILINEATUM DC.		SHRUB	2500	
HELICHRYSUM TRILINEATUM DC.			2500	
HIRPICIUM ARMERIOIDES (DC.) ROESSL.			2500	
MACOWANIA SORORIS COMPTON	SOIL	SHRUB	2500	
MACOWANIA SORORIS COMPTON	SOIL	SHRUB	2500	
NIDORELLA AURICULATA DC.			2500	
OSTEOSPERMUM JUCUNDUM (PHILL.) NORL.			2500	
OSTEOSPERMUM THODEI MARKOTTER		SHRUB	2500	
PENTZIA COOPERI HARV.			2500	
PENTZIA COOPERI HARV.			2500	
PTERONIA SP.	STONY	SHRUB	2500	E

Fig. 5. INQUIRY output. The query was structured in the following way. Dataset: COMPOS. Select: altitudes above 2000 m. Sort: by altitude then by scientific name. Report: Scientific name, substrate, life form, altitude, aspect.

Manual checking is carried out primarily when new specimens are loaded and when a genus is revised. All newly loaded specimens are checked for errors before labels are printed. When a taxonomic revision is published, the specimens in the herbarium are checked and, where necessary, the names are changed in PRECIS. Other errors are corrected as they are encountered.

Two essential items of information about each specimen are contained in the databank but are not necessarily on each specimen. These are the PRENO, which is the unique computer specimen number used to locate the specimen in the system, and the grid reference, which is necessary for accurately mapping the location of the specimen. In order that they may be available on each specimen, this information, plus the collector and number, are printed on gummed labels that may be quickly affixed to the specimen. The SPECTAGs for a genus are sorted by collector and number, so that the tag for a specimen is quickly found.

PROBLEMS AND SOLUTIONS

The new PRECIS system serves the needs of the herbarium and of flora researchers, because its output appears in an acceptable format, because the turnaround time is rapid (overnight in the case of many INQUIRY submissions), and because the cost of operating the system is within the means of the Institute. However, there are a number of problems that still limit the usefulness of PRECIS. These are discussed below in some detail, so that the kinds of problems found in operating a large system with a herbarium context are on record.

1. Inaccuracies in Specimen Data

Three items of information about each specimen are indispensible: its collector, its identification and its locality. In PRECIS all these data are recorded by codes, the collector by the COLREGNO, the identification by the GENSPEC and the locality by the GRID. Unfortunately, numeric codes are subject to errors in encoding and in punching because the characters are easy to transpose and difficult to proof-read. The widespread occurrence of errors in these fields in PRECIS is a major impediment to the usefulness of the system. In order to correct the mistakes, it will be necessary to check manually the data for each specimen in the herbarium and in PRECIS.

The task is expected to be a long-term undertaking, and the following method will be used:

(1) listing of all specimens in a genus are compared with the specimens actually held in the herbarium, and SPECTAGs are applied to all backlog specimens;

(2) specimens in the herbarium which do not appear on the listing are encoded, with the symbol "A" (= "additional") in the Herbarium Code field;

(3) entries on the computer listing which do not match a specimen in the herbarium are marked in the databank with the symbol "X" (= "wrong") in the Herbarium Code field;

(4) mistakes in collectors, dates and grids are corrected as they are encountered.

2. Problems with Non-specimen Datasets

The COLLECTOR dataset has been studied by botanists to make sure that each collector appears only once, even though there may be considerable variation in the spelling and initials of a name on the herbarium specimens. For example, the name "G. E. Gibbs Russell" originally had 17 different entries in the COLLECTOR dataset, due to variations in spelling, hyphenation and initials. A result of these updates has been that the spelling of the collector on the specimen label (which is often incorrect or abbreviated) no longer agrees with the spelling maintained on the specimen in PRECIS. This may cause confusion to inexperienced curatorial assistants in the herbarium.

The GENSPEC datasets of scientific names with authorities are continuously updated to add new or newly-recorded species for the area, and to enter name changes brought about by taxonomic revisions. As soon as the new system was operational, a large backlog of GENSPEC updates was entered to bridge the gap between PRECIS I and PRECIS II. A check-list of all southern African plants is in preparation by the herbarium staff on the basis of the GENSPEC datasets.

3. Lingering Problems from PRECIS I

When the original encoding took place, the computer specimen number (PRENO) that was stamped on the encoding form for each specimen was not also recorded on the specimen itself. Because of this serious oversight, it was impossible for individual specimens to be updated for name changes or other corrections. The problem is being solved by the use of SPECTAGs which are attached to the specimens, so that the PRENO is readily available if the specimen must be updated. All new specimens encoded now have the PRENO stamped on their labels.

After the backlog for PRECIS was encoded, plant collectors at the Institute recorded their collection data on encoding forms, as shown by Morris and Glen (1978). For the convenience of collectors, the full locality and habitat data were filled in only on the first specimen from a site. Unfortunately, this practice for PRECIS was not compatible with the procedure for naming specimens in the herbarium. Collections with the same data were separated to the different herbarium wings to be identified, and often the single form bearing locality information was discarded or returned to the collector, leaving the other specimens without this data. A program has been written to convert the data on these old encoding forms so that they can be loaded into the system.

4. PRECIS – Herbarium Relationships

Several of the problems with PRECIS I were the result of imperfect liaison between the developers and the staff of the herbarium. This could easily have led to the downfall of the entire project. Fortunately, one member of the staff began to unravel some of these problems, and his work was the basis for the restructured system now known as PRECIS II.

After PRECIS II became operational, fresh difficulties arose in fitting computerization requirements to traditional herbarium procedures. Workflow patterns for both PRECIS II and the herbarium have been studied together and incorporated into a single integrated plan for specimen handling. Herbarium personnel have contributed to these plans, and each staff member is kept up to date about changes in procedures. As a result, PRECIS and the herbarium can no longer be regarded as separate. PRECIS is an image of the herbarium, and curatorial changes are not complete until they are incorporated in both.

NEW PROSPECTS

In the eight months that the new system has been in operation, a number of queries have been made for which information was sorted in ways that could never have been contemplated without a computer. Since this had been predicted for PRECIS from the beginning (Morris, 1974), it is satisfying that reality meets the original expectations. The following are examples of results that have been obtained from PRECIS II:

(1) Gunn and Codd (1981) have published an account of botanical collectors in southern Africa. To extend the knowledge of lesser-known and younger collectors, they have extracted additional names from the collector list.

(2) Flowering phenology correlated to grid reference has been determined for about 40 important veld grassses.

(3) The number of specimens known for each whole degree grid square, reported by the SPECIES PER GRID program, has been mapped to show the intensity of collecting over the entire Flora area. This will serve as a guide when planning the Institute's field work in the future.

(4) Lists of grasses collected in each of the eight territories of the Flora area have been compiled. These will be the basis for guides to the grasses in each of these regions.

(5) Distribution maps for species in the genus *Alloteropsis* have been plotted for the *Flora of Southern Africa*. The grids as listed by PRECIS were entered in a data file on the Institute's Hewlett-Packard 9845B and the maps produced by the usual mapping routine.

CONCLUSION

The restructured form of PRECIS is giving the results required by herbarium curators and flora researchers at the Botanical Research Institute. Furthermore, the increasing number of queries submitted by botanists outside the Institute shows that PRECIS will play a significant role in advancing taxonomic work throughout southern Africa. The success of the new system can be ascribed to the following factors:

(1) the simplified databank is quick to operate and its use is relatively inexpensive;
(2) the specimen encoding form is designed to ensure that all data are attached to each specimen, and that locality data are available to herbarium staff when specimens are being named;
(3) the specimen labels are in a format suitable for herbarium use;

(4) PRECIS and herbarium workflows have been planned to complement each other, so that PRECIS continues to grow in parallel with the herbarium.

ACKNOWLEDGEMENTS

Three individuals have been of outstanding importance in the ten-year history of PRECIS: Dr J. W. Morris, Dr B. de Winter and Miss B. Young. Dr H. F. Glen, Dr. R. E. Magill and Mrs J. H. Jooste have also contributed much to PRECIS.

REFERENCES

Edwards, D. and Leistner, O. A. (1971). A degree reference system for citing biological records in southern Africa. *Mitt. bot. StSamml. Münch.* **10**, 501–509.

Gunn, M. and Codd, L. E. (1981). " Botanical exploration of southern Africa". Balkema, Cape Town.

Leistner, O. A. and J. W. Morris. (1976). Southern African place names. *Annals Cape Prov. Mus.* 12, 1–565.

Magill, R. E., Gibbs Russell, G. E., Morris, J. W. and Gonsalves, P. (In press). PRECIS – the Botanical Research Institute herbarium data bank. *Bothalia.*

Morris, J. W. (1973). "Overseas advances in quantitative ecology and computerized data banking in biology". Unpublished report to the Department of Agricultural Technical Services, Pretoria.

Morris, J. W. (1974). Progress in the computerization of herbarium procedures. *Bothalia* 11, 349–353.

Morris, J. W. (1980). Encoding the National Herbarium (PRE) for computerized information retrieval. *Bothalia* 13, 149–160.

Morris, J. W. and Glen, H. F. (1978). PRECIS, the National Herbarium of South Africa (PRE) computerized information system. *Taxon* 27, 449–462.

Morris, J. W. and Leistner, O. A. (1971). Index of localities for southern Africa. *Mitt. bot. StSamml. Münch.* 10, 498–500.

Morris, J. W. and Leistner, O. A. (1975a). Computerization of the National Herbarium, Pretoria. *Taxon* 24, 270.

Morris, J. W. and Leistner, O. A. (1975b). Progress with the computerization of the National Herbarium, Pretoria. *Boissiera* 24, 411–413.

Morris, J. W. and Manders, R. (1981). Information available within the PRECIS data bank of the National Herbarium, Pretoria, with examples of uses to which it may be put. *Bothalia* 13, 473–485.

13 A Review of Herbarium Catalogues

R. J. PANKHURST

Botany Department, British Museum (Natural History), London SW7 5BD, UK

Abstract: During the last 20 years, databases have been created in order to catalogue herbaria in many parts of the world. Some progress has been made in the organization of such projects and the design of the databases, and certain shortcomings have occurred more than once. Significant databases are compared and reviewed.

INTRODUCTION

Why should one want a catalogue to a herbarium? Herbaria are usually organized physically in systematic order, i.e. in a hierarchical scheme of families, genera and species. There is some variation as to which classification is adopted at each level of the hierarchy but, in general, one is expected to find any specimen by looking it up first of all by its scientific name. There is often a secondary ordering by geographical localities within the categories of genus or species. This is the best arrangement for floristic work, involving all the plants of a certain family or genus from one region. It is nearly as good for monographic work, where one is just concerned with plants that are closely related to each other. But what happens if you want to find all the specimens of a certain collector, or those of a certain period, or those which are annotated with information about uses? The only possibility is to search the entire collection, which is usually impractical. In other words, the collection is normally only self-indexed,

Systematics Association Special Volume No. 26, "Databases in Systematics", edited by R. Allkin and F. A. Bisby, 1984, pp. 155–164. Academic Press, London and Orlando.

DATABASES IN SYSTEMATICS
ISBN 0 12 053040 6

and in only one or two ways. To search it in any other way, or combination of ways, a multi-way index is needed, and the only practical way to do this is with a computerized database. Much useful information is contained in conventional herbaria which cannot be accessed in any practical way; as Shetler has said, a herbarium is often not so much a databank but a data crypt.

Existing herbarium databases have produced a variety of useful products. One may cite the Flora of Veracruz (Gómez-Pompa and Nevling, 1973), which produced a check-list of the species in the region. This check-list was invaluable because there was no existing Flora of the area, and workers in the project had no published source to which they could refer. In several cases computers were used to map species distributions from distribution data in databases, such as in the Flora of Warwickshire (Cadbury *et al.*, 1971). Among the many products of the PRECIS system at Pretoria are quoted the study of flowering times (phenology) and the reconstruction of a collector's route (Morris and Manders, 1981). Other possibilities are the retrieval of data about the uses of plants, especially medical properties, and the use of the database for administering loans of specimens.

The following discussion is based on databases for herbaria, but it is likely that the content of these remarks would be much the same if zoological collections were being discussed. One possible difference might be the need to give specimens of disparate sizes a code for their physical location. To give due credit, the largest current system (December 1982) is that in the Department of Zoology, Smithsonian Institution, Washington D.C., which has about 2 million records.

EXAMPLE SYSTEMS

In order to limit the survey to a reasonably small set of examples, only databases with more than 50 000 records (specimens) have been included. This gives us nine different systems to compare at the time of writing (December 1982). Databases in use for botanic garden living collections, of which several exist, are not included. Small databases do not hold as much interest because the problems of planning and management are insignificant, especially if they can all be handled by one person.

The systems listed in Table 1 are largely situated in the USA, Latin America and in British Commonwealth countries. Further details of

Table 1. Comparison of herbarium databases.

Herbarium	*Country*	*No. of specimens*	*Start date, approximate*	*Computer(s)*	*Software*	*Language(s)*	*References*
Notre Dame (Herb. Greene)	USA	65 000	1968	IBM	Local	PL/I	Crovello (1972)
Smithsonian Institution	USA	53 000 (algae) 70 000 types (incomplete)	1969	Honeywell	SELGEM	COBOL	Creighton *et al.* (1971)
Flora of Veracruz	Mexico	60 000	1970	Burroughs, VAX	Local, TESAURO	ALGOL, PL/I	Gómez-Pompa and Butanda (1973)
Brisbane	Australia	360 000	1973	UNIVAC, PDP11	Local	?	
Minnesota	USA	60 000	1973	CDC	System 2000	?	Wetmore (1979)
Bogotá	Colombia	120 000	1974	IBM	Local	COBOL	Forero and Pereira (1976)
Pretoria	South Africa	620 000	1974	Burroughs	PRECIS I, INQUIRY	COBOL	Morris and Glen (1978), Gibbs Russell and Gonsalves (Chapter 12, this volume)
Programa Flora	Brazil	220 000 (500 000)*	1976	IBM	TAXIR	?	Teixeira (1977), Brill (1978)
Mexico City UNAM	Mexico	60 000 (400 000)*	1982	Burroughs	TESAURO, DMS2	PL/I	

* Figure in brackets is for final size when completed.

those in the USA are given by Sarasan and Neuner (1983). None is reported from Japan or the Soviet Union, but this is not to say that none exists. There are no systems of significant size in Europe, according to the above terms of reference. There is perhaps a tendency for the relevant institutions to be fairly young, or to be in a stage of rapid growth. An important criterion may be the ratio of the number of specimens to the number of species. Where this tends to be high, as in European herbaria, the effort required to capture the backlog of data is relatively greater, and this reduces the incentive to use computers (Brenan, 1974). It is also noticeable that none of the systems is run on a commercial basis, maybe because this would inhibit potential users.

CRITICISMS

This section comments on the mistakes made in past projects for the creation of herbarium databases. Since the discussion mainly concerns matters of design and management, the points which are made are necessarily controversial. The intention is to provide some answers to such questions as: "If one wants to set up a herbarium database today, what pitfalls ought to be avoided?"

1. Fixed Format

In early database management programs, users were often expected to present their data in fixed positions on input. This was because of the use of cards for data input, which have now largely been superseded. More modern programs allow the user to prepare data in more flexible ways, and make the computer work harder to interpret the data which it is given. Fixed-format input should be avoided because it is burdensome. Database management programs may only permit the use of fields, e.g. species name or locality, which are fixed in length. This is inconvenient to the user when new data comes along that is too long for the space allowed for it, even though the data may itself be perfectly legitimate. This can only be accommodated by using abbreviations or other mutilations of the data. There is no need for this because computers can be programmed to accept data of variable size and not to waste storage on empty or short fields. Hence programs which can accept data of variable length without wastage are to be preferred.

2. Questionnaires

In nearly all the projects reported so far, a printed questionnaire has been prepared, e.g. for the Flora of Veracruz, in Gómez-Pompa and Nevling (1973), for the Programa Flora in Teixeira (1977) and for the PRECIS system, in Morris and Glen (1978). This would have spaces for specific name, collector's name and so forth, and be filled in by a worker in the herbarium. It would then be given to a data preparation operator or service in some other location, and punched or typed on a keyboard for input to the computer.

A procedure of this kind means that the data are effectively prepared twice, once in the herbarium and a second time in the punch room. The assertion here is that the effort, labour and cost of each process is roughly equivalent, since the same sequence of words is being copied each time, whether in handwriting or on a keyboard. Furthermore, there are opportunities for making mistakes in both places; first when specimen labels are copied (and misread) in the herbarium and again in the data preparation room when the forms are read by someone else.

It would be much better to install the necessary data preparation equipment in the herbarium itself, and have the first line of workers collect the data in machine-readable form straight away. A set of display terminals is needed which may or may not be attached by cable to a more distant computer. There are various technical possibilities, such as time-sharing on a mainframe or minicomputer with remote terminals, or terminals attached to a computer network, or individual microcomputers. In the latter case, data could be transferred on floppy discs for processing elsewhere.

In practice, most systems have used the questionnaire approach. In the past this may have been because the necessary terminals were not available or because there was no accessible computer. There have also been difficulties in finding space for new equipment in cramped conditions, or administrative problems in obtaining permission to use it. Once upon a time, terminals would have seemed too expensive, but this is no longer true. In the author's own institution each terminal would have paid for itself at current prices and data preparation charges when only 10 000 specimens had been processed.

There has also been a tendency to try and score too many fields for each specimen. This is probably because there is a great deal of information that may appear on herbarium labels which research workers would like access to, and so provision is made to record it. There is also the feeling that since the whole collection is going to be worked through, which is in itself a laborious business, one might as well collect all the data in one pass. Against this must be set the fact that some items of data are rare or of low utility, or both. In practice, it seems better to compromise by not attempting to record all information, but just the most valuable part of it. At Pretoria, the PRECIS project began with about 60 items and this has now been reduced to about 30 (Gibbs Russell and Gonsalves, Chapter 12, this volume), which is about the same number as used in the Flora of Veracruz. A minimum set of fields which should be stored for any herbarium label is suggested in Brenan *et al.* (1975). Another way to compromise is simply to record the presence or absence of certain kinds of information, e.g. medical uses, so that somebody looking for such data subsequently can be given a computer print-out to speed searching in the collection.

3. Coding

It is normal practice when entering factual data in a computer to represent certain natural-language words by code numbers which the computer can handle more easily. For example, in the PRECIS project, different life forms such as tree, shrub etc. were coded with the numerals 1, 2, etc., and similarly for other fields. In PRECIS the substitution of codes for words was originally done manually, and it is now clear that this was bad practice. Human operators often commit errors when looking up number or letter codes when preparing data, and mistakes in coded form are harder to detect during proof-reading afterwards, so an unacceptable number of errors can enter the database without being detected. A much better method is to punch all textual data without coding it and then use the computer to recognize code words or symbols afterwards. It may be necessary to put quotation marks or some other markers around important words when only part of a text is to be coded, but it seems that the number of errors avoided by this means is well worth the extra typing. The TESAURO system used in Mexico has this desirable feature. When for example the name of a species is met in the input data, a program checks whether the name is one which has already been seen or not. If it is apparently a new name, then the program prints it and queries whether it really is new. If it is merely a mis-spelling of another name, then the correction can be made at once, and errors of this kind will not enter the database.

To summarize, coding of data is in itself a perfectly acceptable practice, but it is important to carry out the coding by using the computer and not manually.

4. Annotation of Specimens

Continuing the argument in the above section, it is natural to think of annotating the specimens in the collection in various ways as data capture proceeds. The obvious example is to stop and identify unnamed specimens, or to check the identification of some already named. It may also be desirable to add geographical coordinates to labels that lack them. In areas for which gazetteers are not available it is feasible to create a gazetteer from the completed database by sorting the locality field and adding coordinates, as was done with PRECIS and for the Flora of Veracruz. Annotation can be very time consuming and greatly multiply the time required for data capture. It is perhaps best therefore to not annotate any specimens during data capture, or to limit such activities severely.

5. Sequential Record Processing

This is a problem which has arisen with the programming of several of the earlier databases. Programs which coped adequately with a limited number of records (up to 10 000 say) proved to be impossibly slow when searching larger databases. This difficulty arose with earlier programs for the Flora of Veracruz, with PRECIS I, and has been reported with SELGEM. On the other hand, careful attention to the indexing of records and to rapid data transfer on storage devices seems to have avoided such problems with TAXIR and with TESAURO. Managers of new projects should ensure that their software is able to process large numbers of records sufficiently rapidly.

6. Management

According to Sarasan (1981) a large proportion of computing projects in museums in the USA are failures. The main reason given for this is bad management. In more detail, the problems include:

(1) lack of planning, especially in not considering how data is to be retrieved from the database at a later date, and failure to recognize when a database will change with time, so that substantial rates of error correction and addition of new data are not allowed for;
(2) attempts to enter too much information into the database lead to the project taking too long;
(3) human factors i.e. the lack of familiarity of staff with computing technology and failure to acquire new skills;
(4) recruitment of inappropriate staff, stemming from lack of realization that there are different kinds of computing expertise.

It should be clearly realized that projects can fail merely because of management considerations even when they are perfectly viable from the technical point of view. These two aspects should not be confused.

SOFTWARE SYSTEMS

In this section some general comments will be given about the database management programs that have been used. The choice which faces planners for a new project is: to use the existing programs which come with the local computer (if any); to import a package from elsewhere; or to write special programs. Table I shows that the last option has been the one most commonly chosen. This may have been because the systems which are available for this purpose, such as SELGEM and TAXIR, were thought unsuitable, or could not be easily or efficiently implemented on the local computer; or simply that it is more exciting to write the programs for oneself.

It is argued elsewhere (Barron, Chapter 3, this volume) that the best type of database system to use is a relational database system. Individual systems may not be easy to categorize but among those listed in Table I, TESAURO is definitely a relational system. Both versions of PRECIS appear to be relational (Gibbs Russell and Gonsalves, Chapter 12, this volume). TAXIR is intended for sets of records which are "flat", i.e. of a simple non-hierarchical structure, (Brill and Estabrook, Chapter 5, this volume). SELGEM is a hierarchical system with a limited number of hierarchical levels.

Once the backlog of data on specimens in a collection has been captured, the production of labels for new specimens is very conveniently carried out simultaneously with the incorporation of the new data into the database. Indeed, if this is done with suitable display-editing techniques, so that information which appears repeatedly on successive labels is not retyped, there may be considerable labour saving. Probably most of the database projects reviewed here are already doing this. Some are also exploiting modern word-processing equipment to produce high-quality labels, and at least one special-purpose microcomputer program has been written (Pankhurst, 1983).

Advice to leaders of new projects is given by Pettitt (1980), which is recommended except for the remarks on the use of preprinted forms or questionnaires (see above). Much practical and useful detail on how to manage the collection of data is given by Morris (1980). See also Sarasan (1981) for a commentary on management in general, and Sarasan and Neuner (1983) for much useful detail on planning and systems design.

CONCLUSIONS

There are clear advantages in having a herbarium catalogued by computer. These benefits are more clearly seen in smaller collections which are intensively used. If the directors and managers of collections do not agree with this at present, it may be because;

(1) previous projects have tended to fail, although for managerial rather than technical reasons;
(2) there is still a relative lack of computing skills among botanists and museum staff, which hampers the appreciation of such projects;
(3) the costs of such projects have been high – the costs of computing equipment continue to fall, and software becomes more readily available. There will always be a heavy demand on staff time, although proper management can reduce this.

These remarks can perhaps be summarized by pointing to the severe contrast between the time scales of botanical research in herbaria and computing development. The former is very lengthy (a hundred years or more) while either hardware or software often last for as little as five years.

REFERENCES

Brenan, J. P. M. (1974). International conference on the use of electronic data processing in major European plant taxonomic collections. *Taxon* 23, 101–107.

Brenan, J. P. M., Franks, J. W., Raynal, J. and Cullen, J. (1975). Report of working party on electronic data processing in major European plant taxonomic collections. *Adansonia*, ser. 2, **15**, 7–24.

Brill, R. C. (1978). "The TAXIR Primer, MTS version". 3rd edition. Computing Center, University of Michigan, Ann Arbor.

Cadbury, D. A., Hawkes, J. G. and Readett, R. C. (1971). "A computer-mapped Flora. A study of the county of Warwickshire". Academic Press, London and Orlando.

Creighton, R. A. and Crockett, J. J. (1971). "SELGEM: a system for collection management", Information Systems Innovations Vol. 2, No. 3. Smithsonian Institution, Washington, D.C.

Crovello, T. J. (1972). Computerization of specimen data from the Edward Lee Green Herbarium (ND-G) at Notre Dame. *Brittonia* **24**, 131–141.

Forero, E. and Pereira, F. J. (1976). EDP-IR in the National Herbarium of Columbia. *Taxon* **25**, 85–94.

Gómez-Pompa, A. and Butanda, A. (Eds) (1973). "El uso de computadoras en la Flora de Veracruz". Instituto de Biologia, U.N.A.M. Mexico City.

Gómez-Pompa, A. and Nevling, L. I. (1973). The use of electronic data processing methods in the Flora of Veracruz program. *Contr. Gray Herb.* **263**, 49–64.

Morris, J. W. (1980). Encoding the National Herbarium (PRE) for computerised information retrieval. *Bothalia* **13**, 149–160.

Morris, J. W. and Glen, H. F. (1978). PRECIS, the national herbarium of South Africa (PRE) computerised information system. *Taxon* **27**, 449–462.

Morris, J. W. and Manders, R. (1981). Information available within the PRECIS data bank of the National Herbarium, Pretoria, with examples of uses to which it may be put. *Bothalia* **13**, 473–485.

Pankhurst, R. J. (1983). The preparation of collection labels with a microcomputer. *Curator* **26** (4).

Pettit, C. (1980). Factors in the management of museum computer systems. *In* "Information handling in museums" (O. Orna and C. Pettitt, Eds), pp. 88–128. Clive Bingley, London.

Sarasan, L. (1981). Why museum computer projects fail. *Museum News* **59**, 40–49.

Sarasan, L. and Neuner, A. M. (1983). "Museum collections and computers". Association of Systematics Collections, Lawrence, Kansas.

Teixeira, A. R. (Ed.) (1977) "Programa Flora, manual de preenchimento". CNPq, Brasilia.

Wetmore, C. M. (1979). Herbarium computerization at the University of Minnesota. *Syst. Bot.* **4**, 339–350.

14 Flora of Veracruz: Progress and Prospects

A. GÓMEZ-POMPA, N. P. MORENO, L. GAMA and V. SOSA

Instituto Nacional de Investigaciones sobre Recursos Bióticos, Xalapa, Ver., Mexico

and

R. ALLKIN

Biology Department, The University, Southampton SO9 5NH, UK

Abstract: The progress made in the Flora of Veracruz project in the application of databases in its floristic study of the Mexican state of Veracruz is outlined. Major developments include: (1) implementation of a relational system to handle the curatorial database of over 60 000 entries; (2) initiation of the publication of a Flora, with certain bases (i.e. standard format and glossary of terms) laid for a future generalized information system; and (3) current investigation on the application of descriptive database systems. The Flora of Veracruz may be considered unique in that it is located in an area that is rapidly suffering extreme modification of the vegetation, it has stressed the role of databases since the initiation of its activities and it carries out all its operations in a developing country, in spite of the limitations that this may imply.

INTRODUCTION

The Flora of Veracruz has employed electronic data-processing methods since its initiation in the mid-1960's, when the project was

Systematics Association Special Volume No. 26, "Databases in Systematics", edited by R. Allkin and F. A. Bisby, 1984, pp. 165–174. Academic Press, London and Orlando.

DATABASES IN SYSTEMATICS
ISBN 0 12 053040 6

one of the pioneers in the use of curatorial databases for the processing of label information from plant specimens (Gómez-Pompa and Nevling, 1973). The project was initiated by Gómez-Pompa (then at the National Autonomous University of Mexico, UNAM) and Nevling (then at Harvard University) prior to the Flora North America program (FNA), which it helped inspire. It shared the prevailing optimism surrounding activities related to the "automation" of Floras, partly created through interest in FNA.

Since then, the Flora of Veracruz has consolidated and expanded its activities. One of its major objectives continues to be the investigation and application of databases in its long-term study of the flora of this Mexican state. Veracruz is located on the Gulf coast, between 17 and 23°N. The altitude varies from sea level up to over 3000 m above sea level at the Pico de Orizaba. The vegetation is extremely varied and consists of various types of evergreen or deciduous "selvas", and forests, grasslands, palm stands, mangroves, coastal dunes and even certain types of arid or semi-arid vegetation (Gómez-Pompa, 1973). The principal agricultural activities are related to coffee, sugarcane and fruit production and to cattle rearing.

In this paper we review progress in the application of database systems in the Flora of Veracruz up to the present, and the outlook for the future. The three principal areas of progress have been:

(1) changes in the organization of the curatorial database and the mechanism for handling specimen label data. Simultaneously new computing facilities were installed;
(2) progress in the production of a Flora;
(3) feasibility studies for a descriptive database.

CHANGES IN THE CURATORIAL DATABASE

Several changes have been made to the organization of the curatorial database system, which in part reflect historical changes in the project itself. The Flora of Veracruz project was initiated at the Institute of Biology at UNAM in Mexico City and the original database system used was centred in the computing facilities there. In 1976, the project's base was shifted to the National Institute for Investigation of Biotic Resources (INIREB), then a fledgling research institution in Xalapa, Veracruz.

The transfer to Xalapa has had far-reaching effects on the direction of research. First, a new herbarium had to be created. The project could no longer count on the existence of a relatively large, established herbarium, such as that in the Institute of Biology, for identification and other activities. The project staff itself was a second major factor. Most of the experienced investigators remained in Mexico City and the new herbarium was founded using mostly undergraduate students with little or no experience in either routine taxonomic activities, such as identification, or the more sophisticated aspects of floristic research. Finally, the database system itself was becoming obsolete. Its applications were limited and the recovery of information was exceedingly slow. This was further complicated by the physical separation of the computer in Mexico City from the herbarium in Xalapa during the first few years after the move, requiring the use of a cumbersome system of triplicate labels for the initial recording of information and to provide temporary specimen labels. The separation also caused considerable delay in the production of labels and the reception of results from any consultation of the databank.

The problems relating to the curatorial database have been diminished by two major changes in our operations. First, a Digital VAX 11/780 computer has been installed in our new building in Xalapa. Secondly, a new relational database system has been implemented to streamline our activities. The new system permits the direct capture of information from specimens via terminals, and thus eliminates the need for triplicate labels. It also handles on-line consultations and is considerably more flexible and rapid than the earlier system.

We have yet to remedy certain minor problems related to the quality of the herbarium labels produced. We have been unable to obtain a suitable acid-free paper for labels within the country, and have discovered that none of the standard carbon-free inks used in computer printing are sufficiently durable to last the life of a herbarium specimen. We hope to remedy the printing problem through the substitution of a letter quality printer, but cannot improve the paper since a suitable quality is not produced in Mexico.

Even though we have operated a curatorial database for a number of years, it has not yet been fully accepted by a significant portion of the Mexican botanical community. This is illustrated by the very small number of consultations solicited by Mexican investigators not associated with the Flora of Veracruz, as compared to many more requests from outside the country. Gradually, however, as electronic data processing methods have become more common, more interest has been expressed in the use of computerized methods. For example, a second Mexican herbarium, the National Herbarium in the previously mentioned Institute of Biology, is now beginning to build a curatorial database.

PROGRESS IN PRODUCTION OF A FLORA

Despite the slowly growing acceptance of the database, only very few investigators consider a database in itself to be a valid product of research. In Mexico, investigation continues to be gauged more by the number of publications produced, than by any other result or application. Thus, very early in the development of the Flora of Veracruz project, the decision was made to initiate the preparation of a standard published Flora. This would satisfy the requirements of fund-granting agencies, as well as forward investigation of the plant groups present. Obviously, when the first treatments were being prepared in the early 1970's, it was even more difficult than now to produce them by automated or even partially automated means. Thus, in spite of the underlying interest in the applications of databases to taxonomy, the treatments of the individual families were initiated using conventional methods. The help of experts was enlisted and the Flora of Veracruz staff itself also embarked on the preparation of the descriptions of some families. Prior to tackling larger taxa, most of our novice taxonomists initiated their work in this area with the treatment of relatively small, uncomplicated families.

In the Flora of Veracruz series, each family is published in a separate monograph of fascicle, of which about eight are issued per year. Each fascicle contains as a minimum dichotomous keys, synonyms and type citations, at least for names originally described from Veracruz. It also includes common names, detailed morphological descriptions, specimens examined, distribution maps, line drawings, uses, and ecological and phenological information. The series is edited in Spanish.

FUTURE IMPLEMENTATION OF A DESCRIPTIVE DATABASE

However, even as plans were made to produce and publish the Flora of Veracruz by conventional mechanisms, steps were taken in anticipation of a possible future implementation of automatic methods for taxonomic description. A standard format was created to be followed in the description of any taxon in the fascicles (Gómez-Pompa *et al.*, 1978). This was directed towards providing uniform and reasonably complete descriptions, and also at trying to ensure at least minimal levels of comparability for the possible future storage or manipulation of the information in database systems.

1. Glossary

Similarly, steps were taken to provide for the standardization of the botanical terminology used in the project. This was aimed both at standardizing the use of taxonomic terms in the fascicles, as well as coordinating and standardizing any use of the same terminology in a database system. We have accomplished this through the careful preparation of a glossary with definitions and English equivalents for the most important taxonomic terms in Spanish. A final version containing approximately 1900 entries, and 950 illustrations is presently in press (Moreno, 1983). Many manuscripts for the Flora of Veracruz are received in English and must be translated into Spanish by the editors without any alteration in meaning.

As an extension of the original glossary, a database containing taxonomic terms and definitions in six languages (Spanish, English, Latin, French, German and Portuguese) is also being created. This on-line system will be especially useful to students and inexperienced personnel for the correct interpretation and application of taxonomic terms. It may also be valuable as a future complement to a descriptive database.

2. Fundamental Structural Problems

These two measures, the standard format and the glossary, have not solved all the problems related to the production of uniform descriptions. Certain of these problems are inherent to taxonomic terminology in general and will not be resolved. In particular, the partial overlap in meaning between descriptive terms or the inclusion of completely different features in the definitions of supposedly comparable terms will not be eliminated by the standardization of the definitions alone (for example, "cannose" refers to colour and "hirsute" to hair length and stiffness, although both are used to describe vestiture). On the other hand, problems concerning the contributors to the Flora might be resolved. For example, many taxonomists outside the project at present ignore our proposed description order and follow the one most familiar to them. Minor discrepancies are corrected by the editors, but the entire rewriting of descriptions is not practical and may offend the authors. Similarly, we can never be absolutely certain whether the terminology has been interpreted or applied as we suggest.

3. Identification Aids and Descriptive Data

The relative inexperience of many of the staff members and the limited size of the herbarium have continued to pose difficulties for the day-to-day operation of the project, for example, in the routine identification of specimens. Thus, beginning six years ago, work was initiated with Pankhurst, implementing his suite of programs for automatic identification and description (e.g. Pankhurst, 1975). Allkin developed a character set and data matrix to be used with Pankhurst's programs for the description of the Angiosperm families present in Veracruz (Allkin, 1981). A family identification system was given priority for several reasons. Identification to family is often the most difficult and time-consuming step for an inexperienced worker. In our situation, the lack of experience was compounded by the absence of an established plant collection that could be used to confirm tentative determinations. A computer based system, in particular, was suitable to our needs for two other reasons. Much of Veracruz is inadequately explored and species listings must continually be revised. For example, it is not uncommon to receive first reports of even new genera for the state. Occasionally, genera new to science are still found, such as *Olmeca* in the woody bamboos or an as-yet undescribed new genus of orchids. Consequently, any system capable of incorporating new information without substantial modification or republication is extremely practical. Secondly, the difficulties which may be encountered in the identification of specimens belonging to new or poorly understood species, as is the case with many tropical taxa, can be alleviated somewhat by a flexible identification system that provides several different options (i.e. matching, on-line, etc.) to be chosen by the user. Furthermore, other projects in our Institute frequently need specialized identification aids, which can be prepared relatively simply by automatic methods.

For example, a special key to the woody families of Veracruz is particularly useful to our investigators studying tropical woods.

The Flora of Veracruz family-key system may be used to produce dichotomous keys or polyclaves, or be consulted by matching or on-line techniques.

An interesting offshoot of this activity using programs designed to produce descriptions in English, has been the need to adapt the programs to Spanish and to design characters that are logical and understandable in that language. Even in the simplified format of computer-produced descriptions, the changes in syntax for Spanish involve more than the simple transposition of adjectives, and are complicated by the generally less suitable structure of the Spanish language for taxonomic description. Spanish versions now exist for the entire group of programs.

An alternative description generation program CONFOR (Dallwitz, Chapter 23, this volume), has recently been acquired to give the project greater flexibility. A program is currently being written to convert data from one format to the other (Pankhurst, pers. comm.).

4. Unskilled Scoring and Standardization

Directly related to the development of the family identification database has been another series of recent experiments on the use of similar methods, particularly for description. We are investigating the possible use of a number of relatively inexperienced students or technicians for the observation and scoring of plant attributes in response to a standard set of characters. This is being accomplished using the same suite of programs, an experimental set of vegetative characters (Moreno and Allkin, 1981) and those species in the Clavijero Ecological Park. The park is adjacent to our institute and is both readily accessible and relatively well studied. Any descriptions or identification aids produced have practical applications.

We are particularly interested in discovering whether:

(1) observations carried out by numerous people with varied backgrounds are comparable and consistent, and to what degree;
(2) the types of character formulations that are most easily understood and applied by semi-skilled observers; and
(3) the practicality of carrying out such an enterprise on a large scale.

The large number of students and technicians associated with the project make this type of study feasible.

Preliminary results indicate, for example, that most of the personnel preferred traditional botanical terminology, instead of characters based on the logical representation of plant structures, in spite of their recent initiation to taxonomy; even though a logical series of objective characters may more accurately portray the actual conditions observed. For example, most users preferred to describe leaf shape using one character with ten common terms as states rather than three characters separating different aspects of leaf shape symmetry.

The activities described form part of our search for a convenient way to handle descriptive information. We need a generalized information system to coordinate our activities. Information is currently transferred manually between our database activities and our conventional activities, such as the preparation of the fascicles. We require a system that would coordinate data capture, storage and retrieval and avoid manual transfer of information from one activity to another.

OTHER APPLICATIONS

We have already begun to apply the experience acquired in the Flora of Veracruz in a new project, the EthnoFlora of Yucatan. We are using the same systems and techniques as in Flora of Veracruz, with one major addition. The cultural wealth of the Yucatan Peninsula is such that we are compiling ethonobotanical information about the plant species present.

Another project involving similar techniques is the formation of a databank on the useful plants of Mexico. Over 70 bibliographic sources have been reviewed to provide 70 000 different entries.

A computerized cartographic system involving climatological and geographical information (some of it from satellites) in the interpretation and representation of the correlation between the physical environment and plant distribution in Veracruz, has also been developed.

We have extended our curatorial system and begun to accumulate information on specimens collected throughout the country. We hope gradually to expand our activities and contribute significantly towards the development of a national Flora.

CONCLUDING REMARKS

(1) We are working in a developing country with all the limitations in personnel and infrastructure that this implies.

(2) Veracruz is subject to the increasing destruction or irrevocable modification of many elements of its vegetation. This requires the rapid and efficient registry and study of the constituent species of the flora before their disappearance.

(3) The Flora of Veracruz project has stressed the role of database systems in floristic studies since the initiation of its activities; first, by way of a curatorial database and, more recently, through experimentation with methods for automatic identification and description.

(4) We have been able to achieve many of our goals, in spite of our limitations,

  (a) a curatorial database system with over 60 000 entries from six major herbaria (ENCB, F, A, GH, MEXU and XAL);
  (b) a herbarium with over 75 000 specimens amassed in six years;
  (c) a series of fascicles (32 to date) describing the plant families in Veracruz, with bases laid for a future integrated system; and
  (d) important new activities involving the application of automatic methods for identification and description.

We would also point out that what we have accomplished can be carried out elsewhere. We have found databases valuable. By means of our curatorial database we have access not only to the label information of our own collection, but to that of several other important herbaria with Mexican specimens. We are able to consult this information for specialist studies or as a service to the general public. The family-key system has made identification easier for our semi-trained staff.

However, to further the application of databases in the Flora of Veracruz, we depend upon progress in database research as a whole. Given present circumstances in Xalapa, we are not able to design the general taxonomic data management system which we urgently need to coordinate our activities. Nevertheless, we have overcome most of the problems related to the initiation of a large-scale floristic program and have the experience necessary to operate such a system.

REFERENCES

Allkin, R. (1981). "The computer-assisted production of aids to the identification of the Angiosperm families of Veracruz: An introduction and brief project outline". [Document No. 8130023.] INIREB, Xalapa, Mexico.

Gómez-Pompa, A. (1973). Ecology of the Vegetation of Veracruz. *In* "Vegetation and Vegetation History of Northern Latin America". (Graham, A. Ed.), pp. 73–148. Elsevier, Amsterdam.

Gómez-Pompa, A. and Nevling, L. I., Jr. (1973). The use of electronic data processing methods in the Flora of Veracruz Program. *Contr. Gray Herb.* 203, 49–64.

Gómez-Pompa, A., Nevling, L. I., Jr., Sosa, V. and Moreno, N. P. (1978). Guide for the contributors to the Flora de Veracruz. [Document No. 7930006.] INIREB, Xalapa, Mexico.

Moreno, N. P. (1983). "Glosario Botánico Ilustrado". Instituto Nacional de Investigaciones sobre Recursos Bióticos, Xalapa, Mexico. (In press).

Moreno, N. P. and Allkin, R. (1981). "Reporte preliminar sobre el conjunto de caracteres vegetativos para la Flora de Veracruz". [Document No. 8130067.] INIREB, Xalapa, Mexico.

Pankhurst, R. J. (1975). Identification by Matching. *In* "Biological Identification with Computers", (R. J. Pankhurst, Ed.), pp. 79–91. Academic Press, London and Orlando.

15 | The Vicieae Database: An Experimental Taxonomic Monograph

M. E. ADEY, R. ALLKIN, F. A. BISBY, R. J. WHITE

Biology Department, The University, Southampton SO9 5NH, England

and

T. D. MACFARLANE

Western Australian Herbarium, South Perth, W.A. 6151, Australia

Abstract: The purpose of this project is to construct a descriptive monographic database for species of the tribe Vicieae, to set up an information centre for the tribe, and to explore the needs and reactions of users of the taxonomic products involved in the project. The tribe Vicieae (the vetches and peas) was chosen as it has a conveniently small size (327 species), is well known and well represented in Europe, and is of considerable economic importance to humans. The project has two phases. In the first the Vicieae Database was constructed: it was designed for species, a species list was agreed, and data for species were collated and entered. In the second phase the database was used in an experimental Information Centre: data-enquiry, pamphlet and tailor-made product services were offered on an experimental basis to user panel. No major program development is being undertaken; we have used existing data-handling facilities to establish whether a new database program is needed. We have five database files: nomenclatural, morphological, geographical, curatorial and chemical. EXIR holds the main database for direct retrieval and for passing files to other programs. The program EXIRPOST is used to tabulate the products of retrieval or to prepare them for input into application programs. The program CONFOR is

Systematics Association Special Volume No. 26, "Databases in Systematics", edited by R. Allkin and F. A. Bisby, 1984, pp. 175–188. Academic Press, London and Orlando.

DATABASES IN SYSTEMATICS
ISBN 0 12 053040 6

used to create printed descriptions of the type found in Floras, and the program SYNONYMS produces nomenclatural lists. The development of the Vicieae Database Project, and the services offered by it, are being closely monitored to establish criteria for the development of information centres in the future.

INTRODUCTION

The purposes of the Vicieae Database Project are:
(1) to construct an experimental, descriptive, monographic database for the tribe Vicieae (family Leguminosae);
(2) to set up an Information Centre to collect and disseminate information about the Vicieae, based on the services provided by the Project personnel and the database;
(3) to explore the relationship between the producers and users of taxonomic products using services provided by the Information Centre.

This paper outlines some of the problems in the existing producer/user relationship, describes the construction of the database, and discusses the unconventional taxonomic services provided in the initial phase of the Project and anticipated for the final year.

As mentioned by Bisby (1984; Chapter 2, this volume), many of the problems in the present relationship between the producers and users of taxonomic products arise because conventional taxonomic revisions cannot keep pace with the wealth of data being published. For example, it is difficult to obtain the taxonomic distribution pattern of characters, such as chemical data, within sets of taxa. Such a distribution pattern might be of use to an entomologist interested in a list of Vicieae species with a particular amino acid in their leaves, and this could then be correlated with the distribution of aphid species among the Vicieae. Few classifications and identification keys are available for subsets of characters or of taxa, for example, for identifying taxa by their seeds. Conventional revisions also cannot accommodate different classifications of the same set of taxa which occur due to incongruence between the different character sets by which the taxa were classified. The Vicieae Database is being used to explore these problem areas. We are collecting and entering comparable codified descriptions of taxa, updating the database as new data become available, and retrieving subsets of data defined using subsets of characters, taxa, or both. We plan to use such subsets to produce special-purpose classifications or identification aids. The products are aimed at scientists ("users"), who are not necessarily taxonomists; there is no direct link between the users and the database, and all requests for information are handled by the project personnel.

In the first phase of the project the Vicieae Database was constructed: it was designed as a species level database, a preliminary species list was agreed, and data for many species were collated and entered (Bisby *et al.*, 1983). In the second phase, which has just begun, the database is used to provide various novel taxonomic services for the Vicieae to user panel. The panel is composed of plant and crop taxonomists, phytochemists, a seed expert, ethnobotanists and plant breeders (all of whom are interested in the Vicieae) and specialists in taxonomic computer databases. The members of the panel are asked to request the products available and to give candid and constructive assessment afterwards. The members have also been invited to discussion meetings at Southampton. Since its conception, the users panel has expanded its scope to become the Vicieae Database Study Group which combines the initial function of evaluating the Vicieae Database Project with a longer-term function as a forum for exchanges concerning the Vicieae. Already we have had a large amount of support from the Study Group, and have also assumed the role of an information centre for queries and requests about the Vicieae made by scientists not in the Study Group.

The tribe Vicieae (*Lathyrus, Pisum, Vicia, Lens, Vavilovia*) was chosen for four main reasons: it is of a small size (788 taxa; 327 species); it is well known and well represented in Europe; and it is of considerable importance to humans as a source of food, fodder, pharmaceuticals, toxic contaminants, weeds and ornamentals. Members of the project team had previous experience of the taxonomy, morphology and chemistry of tribes in the Leguminosae.

COMPUTER PROGRAMS

As we intended to explore existing software no major program development has been undertaken in the project. This policy has allowed us to evaluate programs already available to establish whether there is a need for a new taxonomist's database management program, and, if so, what its specifications should be. The project has not been entirely automated, and project personnel use their experience to amend or add to computer searches, or to respond where computer data-handling facilities are not appropriate.

The Vicieae Database Project currently operates part of the information service envisaged by Bisby (1984; Chapter 2, this volume). It represents an advance over other monographic databases (Watson and Dallwitz, 1981; Sieg, Chapter 11, this volume) in that the multi-purpose database can be used for tasks in four of the boxes of Bisby's "ideal flow diagram" (see Fig. 1 in Bisby, Chapter 2, this volume) and eventually will be available for tasks in all five boxes. We use a number of different programs which allow us to produce printed descriptions, to retrieve sets of data describing taxa, and to pass the products of retrieval on to programs for key-making and phenetic cluster analysis.

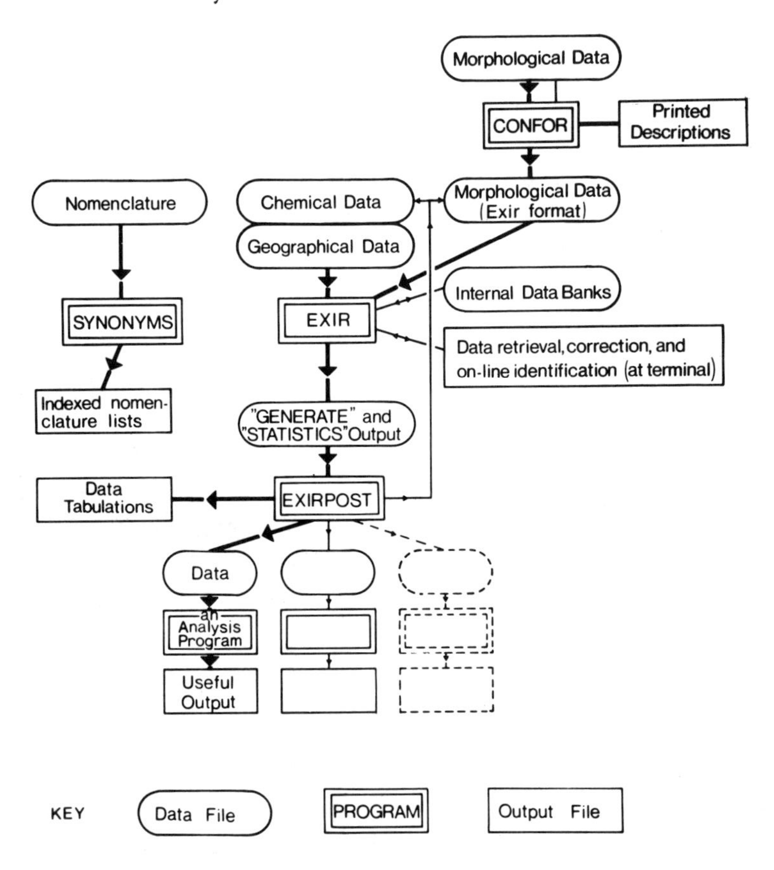

Fig. 1. Flow of information between the programs used in the Vicieae Database Project (from Bisby *et al.*, 1983).

The flow of information between the programs used in the Vicieae Database Project is shown in Fig. 1. The programs SYNONYMS, EXIR, CONFOR, and EXIRPOST run interactively on the University's ICL 2970 computer. The EXIR program (Abbott, 1976) holds the main database for direct retrieval and for passing files containing retrieved data subsets to other programs. The products of on-line retrieval can either be displayed or printed directly, or passed to the program EXIRPOST (written by one of us, R. J. White) to be tabulated with adjustable headings and column widths. All of our printed results can be passed from the ICL 2970 to our Research Machines 380Z microcomputer for text editing before being stored or printed using a Diablo 1650 daisywheel printer. Alternatively, EXIRPOST can rearrange the retrieved data into other formats, such as for input to Pankhurst's key-making programs (Pankhurst, 1976) or to the LINKAGE program (a single-link cluster analysis program by Fleming and Appan, 1972). The program CONFOR (Dallwitz, 1980) is being used to create printed descriptions similar to the type found in Floras. At present, morphological data is entered separately to the EXIR and CONFOR programs as each has facilities not available on the other and each requires the data in a different format. However, the morphological data in both contain the same information, and an interface program between CONFOR and EXIR is being written by Dallwitz. The program SYNONYMS (written by R. J. White) which creates sorted synonymized nomenclatural lists from randomly arranged data files, is also independent of the EXIR program.

DATABASE CONTENT

At present we have five database files (nomenclatural, morphological, geographical, chemical and curatorial) which can be used for on-line retrieval using EXIR. The chemical database file is discussed separately by Babac and Bisby (Chapter 18, this volume). For simplicity and security, additions and deletions to each database are made by only one person, although retrievals are made by any of the Project personnel.

The first steps in the construction of the database files were to coordinate the conventional structure of descriptive data with the structures allowed by the programs selected for handling them, to formulate the character sets, and to devise the wording by which the characters and character states would be entered and retrieved.

The nomenclatural database file contains the accepted names and synonyms of 419 taxa taken from published nomenclatural accounts. Taxonomic revision was not a primary goal of the project but some decisions did become necessary when compiling the nomenclatural list. For example if there were several taxonomic treatments of a group we had to decide which to follow (this problem was particularly severe with common, widespread species). Difficulties also arose when considering species for which there exist extensive, conflicting, infraspecific classifications, and also with poorly known species, particularly for South American plants. Bibliographic details, doubtful synonymies, and "provisional" names of uncertain status are also included in the list. The names are sorted and cross-indexed by the SYNONYMS program, which produces printed lists in several formats. For example, a short list of accepted names can be produced, part of which is shown in Fig. 2 (Allkin *et al.*, 1983a). Alternatively,

115 Lathyrus bijugatus White
116 Lathyrus binatus Pancic
117 Lathyrus blepharicarpus Boiss.
118 Lathyrus boissieri Sirjaev

Fig. 2. A sequence from the list of species and subspecies in the Vicieae (from Allkin *et al.*, 1983a).

a synonymized list might be produced with each synonym entered twice, once in its alphabetical position with reference to its accepted name, and once following its accepted name, as shown in Fig. 3 (Allkin *et al.*, 1983b). Lists have also been requested with spaces left so that handwritten notes can be inserted by gardeners and for the seed collection of the Vicieae Project.

115 Lathyrus bijugatus White
116 Lathyrus binatus Pancic
Lathyrus bithynicus L.
Valid name: 625 Vicia bithynica (L.) L.
117 Lathyrus blepharicarpus Boiss.
118 Lathyrus boissieri Sirjaev
Synonym: Lathyrus nervosus (Boiss.) Boiss. non Lam.
Synonym: Orobus nervosus Boiss.

Fig. 3. A sequence from the synonymized list of species and subspecies in the Vicieae (from Allkin *et al.*, 1983b).

The information contained in the descriptions of the 71 species so far included in the morphological database file resembles that given in the entries for each species in Flora Europaea (Tutin *et al.*, 1964–80); for example, there is information about the shape, size, outline, colour, hair-covering, and arrangement of vegetative, floral and fruit structures (99 taxonomic characters in total). Subspecies, varieties and cultivars have not been included as distinct entries, although we have collected information about the former two categories of taxa.

Creating the morphological character list and collating morphological data took longer than anticipated as a consequence of the demanding standard that we have chosen. The main source of morphological data has been our own observations of herbarium specimens carefully identified using the existing published keys. Our living plant and seed collection has already proved useful for ensuring comparability of data, and more checks for comparability will take place in the coming year. The morphological characters have been selected for their relative ease of recording (important factors being the distinctiveness of the character states and the relative simplicity of the techniques used to determine them) and also their occurrence in the taxonomic literature. For each taxon a large number of specimens are surveyed from the principal herbaria, and from among these a minimum of six (or multiples of six for species in which subspecies are recognized) are carefully selected to incorporate the range of variation within that taxon. These selected specimens, along with specimens from our own herbarium (SPN) and our collection of living plants are submitted to detailed observation of the entire character set.

Morphological observations from specimens and live plants are recorded on a standard score sheet pre-prepared to facilitate comparable recording of information between taxa. It is intended to supplement the morphological records with data taken from reliable references (such as the work of Kupicha and Gunn) to extend the range of variation that we have recorded for each taxon and to add information such as the Stomatal Index (Kupicha, 1974, 1976, 1982) which we have insufficient time to observe ourselves. Diploid chromosome numbers for 149 species have already been collated from the literature, and we are building up a library of taxonomic references, maps, diagrams and photographs from the literature.

The morphological data is prepared in both EXIR format, for the EXIR program, and in DELTA format, for the CONFOR program, in order to take advantage of the differing facilities in the two programs. As both programs can handle a rectangular table of species entries for multistate and quantitative taxonomic characters it would be at least possible to enter identical data to both. However, this would deny ourselves the useful comment and character variability facilities in CONFOR and the facility to make precise retrievals on variable characters possible with EXIR. By structuring the same taxonomic characters in different ways for the two programs we take advantage of both facilities.

Each taxonomic character is entered as a single character in DELTA format (Dallwitz, 1980, 1983) and the resulting data table is used for generating printed descriptions. Two useful facilities are that comments can be attached to any taxon, character or character state for optional inclusion in the printed description, and that taxon variability is handled well. Taxon variability is incorporated by specifying that several states of one character are present, and coding them to be connected by the words "or", "to" or "and". For example, a species may have leaflets described as "ovate or lanceolate". In EXIR format (Abbott, 1976) a multistate taxonomic character is entered as a number of binary EXIR characters, one for each state of the taxonomic character, and a quantitative taxonomic character is entered as two quantitative EXIR characters, a maximum and a minimum. The resulting data table is used for retrievals. EXIR allows precise retrieval on any permutation of EXIR character states but cannot accommodate any taxon variability in a single EXIR character. The splitting of a single taxonomic character into several EXIR characters, as described above, allows the variability to be accommodated, either by recording "present" for more than one of the binary EXIR characters formed from one taxonomic character, or by entering different values for the maximum and minimum quantitative EXIR characters.

Taking the characters for the shape and size of the standard petal as examples (Fig. 4), the shape character has three states: oblong, pandurate and obovate. In DELTA format the shape is treated as a single three-state character (character number 36 in Fig. 4) and both the character and its three states are given explanatory comments in angled brackets for optional printing. For instance, a description might include the phrase "standard ⟨shape⟩ oblong or pandurate".

In EXIR format the shape is treated as three binary characters (character numbers 109–111) and the species that had oblong or pandurate standards would be recorded as oblong (109), pandurate (110) and not obovate (111). We can search the database for species whose standards are sometimes oblong (by searching for: 109, oblong) or whose standards are always oblong (by searching for: 109, oblong and (110, not pandurate) and (111, not obovate)). The

(i) DELTA format

46. standard ⟨shape⟩
    1. oblong ⟨banner and claw of approx. equal width, not or scarcely separated by a waist⟩
    2. pandurate ⟨=platonychioid, banner and claw of approx. equal width and separated by a pronounced waist; standard usually deeply cleft apex⟩
    3. obovate ⟨=stenonychioid, banner wider than claw⟩

47. banner of standard petal/mm long/

(ii) EXIR format

Standard shape 1 (109, CODE, oblong, not oblong),
Standard shape 2 (110, CODE, pandurate, not pandurate),
Standard shape 2 (111, CODE, obovate, not obovate),

Standard banner length min (112, ORDER FROM 1 TO 30 BY 0.1 IN mm)
Standard banner length max (113, ORDER FROM 1 TO 30 BY 0.1 IN mm)

Fig. 4. Examples from the morphological character and character state list used in the Vicieae Database Project, (i) in DELTA format, and (ii) in EXIR format.

standard petal length character is a single DELTA character, number 47, for which a value or range can be inserted for printing, as in "banner of standard petal 5 to 8 mm long". The equivalent is a pair of EXIR quantitative characters which would, for instance, allow the species to be retrieved in a request for those with a minimum standard length less than 7 mm or for those with a maximum standard length greater than 7 mm.

The difficulties in handling descriptive data are discussed elsewhere

Name	An	At	Bo	Bz	Cb	Cl	Cs	Ec	Es	Gf	Gu	Gy	Hd	Ni	Pe	Pg	Pn	Sn	Ur	Vz	Wi
Lathyrus americanus	—	At	—	Bz	Cb	Cl	—	Ec	—	—	—	—	—	—	—	—	—	—	Ur	—	—
Lathyrus berterianus	—	—	—	—	—	Cl	—	—	—	—	—	—	—	—	—	—	—	—	—	—	—
Lathyrus cabrerianus	—	At	—	—	—	Cl	—	—	—	—	—	—	—	—	—	—	—	—	—	—	—
Lathyrus campestris	—	At	—	—	—	Cl	—	—	—	—	—	—	—	—	—	—	—	—	—	—	—
Lathyrus hasslerianus	—	At	—	Bz	—	—	—	—	—	—	—	—	—	—	—	—	—	—	—	—	—
Lathyrus hookeri	—	At	—	—	—	Cl	—	—	—	—	—	—	—	—	—	—	—	—	—	—	—
Lathyrus linearifolius	—	At	—	Bz	—	—	—	—	—	—	—	—	—	—	—	—	—	—	—	—	—
Lathyrus lomanus	—	—	—	—	—	Cl	—	—	—	—	—	—	—	—	—	—	—	—	—	—	—
Lathyrus macropus	—	At	—	—	—	—	—	—	—	—	—	—	—	—	—	—	—	—	—	—	—
Lathyrus macrostachys	—	At	—	Bz	—	—	—	—	—	—	—	—	—	—	—	Pg	—	—	—	—	—
Lathyrus magellanicus	—	At	Bo	—	Cb	Cl	—	Ec	—	—	—	—	—	—	Pe	—	—	—	—	—	—
Lathyrus multiceps	—	At	—	—	—	Cl	—	—	—	—	—	—	—	—	Pe	—	—	—	—		—
Lathyrus nigrivalvis	—	—	—	—	—	—	—	—	—	—	—	—	—	—	—	—	—	—	—	—	—
Lathyrus nitens	—	—	—	Bz	—	—	—	—	—	—	—	—	—	—	—	—	—	—	Ur	—	—
Lathyrus paraguayensis	—	—	—	—	—	—	—	—	—	—	—	—	—	—	—	Pg	—	—	—	—	—
Lathyrus paranensis	—	At	—	Bz	—	—	—	—	—	—	—	—	—	—	—	—	—	—	Ur	—	—
Lathyrus parodii	—	At	—	Bz	—	—	—	—	—	—	—	—	—	—	—	—	—	—	—	—	—
Lathyrus pubescens	—	At	Bo	Bz	—	Cl	—	—	—	—	—	—	—	—	—	Pg	—	—	Ur	—	—
Lathyrus pusillus	—	At	—	Bz	—	Cl	—	—	—	—	—	—	—	—	Pe	Pg	—	—	Ur	—	—
Lathyrus subandinus	—	—	—	—	—	Cl	—	—	—	—	—	—	—	—	—	—	—	—	—	—	—
Lathyrus subulatus	—	At	—	Bz	—	—	—	—	—	—	—	—	—	—	—	—	—	—	Ur	—	—
Lathyrus tomentosus	—	At	—	—	—	—	—	—	—	—	—	—	—	—	—	—	—	—	Ur	—	—
Lathyrus tropicalandinus	—	At	Bo	—	—	—	—	—	—	—	—	—	—	—	Pe	—	—	—	—	—	—

KEY

An	Antilles	Cs	Costa Rica	Hd	Honduras	Ur	Uruguay
At	Argentina	Ec	Ecuador	Ni	Nicaragua	Vz	Venezuela
Bo	Bolivia	Es	El Salvador	Pe	Peru	Wi	West Indies
Bz	Brazil	Gf	French Guiana	Pg	Paraguay		
Cb	Columbia	Gu	Guatemala	Pn	Panama		
Cl	Chile	Gy	Guyana	Sn	Surinam		

Requested by Dr. F. Kupicha. Vicieae Database Project, (C) F. A. Bisby and R. J. White 1983.

taken from the Vicieae Project EXIR database and tabulated

(Allkin, Chapter 22, this volume). The arrangement described above that we have devised for searching the Vicieae data in EXIR format is, as far as we know, the only arrangement reported that succeeds in accommodating taxon variability in a data table so that retrieval can be based on the variable characters themselves. However, this is achieved at the cost of an awkward set of commands for querying the database; the operator must in some cases enter clumsy (and hence error-prone) commands and refer frequently to the expanded EXIR character list. A similar awkwardness arises from the accommodation that we have used for dependent characters.

The geographical database file contains records for 298 species (401 taxa); it contains presence/absence records for both continents and countries and includes a character which gives the source of the record. Unlike the records in the morphological database file, the geographical and nomenclatural records are all taken from the literature. Figure 5 shows an example of a response to a query for a list of *Lathyrus* present in South America, using tabulation facilities provided by the program EXIRPOST.

REVISION OF THE DATABASE

Insertions and deletions of data are made to the database files in batches, and the version number of each file is increased with each batch of changes made. Prior to the initiation of the published products, the batches were entered whenever sufficient new data had accumulated to warrant making a new version of a database file. However, with the introduction of the pamphlet series of products which are sent to users, the batches are entered just prior to the issue of a new set of publications. For example, new versions of the morphological database file will be made in February, May and July. We are monitoring user reaction to assess the feasibility of the different revision rates from the viewpoint of both project personnel and product users.

USING THE DATABASE

So far we have used the database for direct retrieval in reply to query, search and catalogue requests from our Study Group. A query request involves retrieving the data from particular cells in the database; for example, determining the stigma orientation of *Lathyrus* species in section *Notolathyrus*. A search seeks taxa with a particular combination of character states; for example, for taxa with yellow flowers and broad, spathulate stigmas. A catalogue request would be for all or a subset of taxa to be printed as a catalogue, indexed by specified characters with selected data tabulated against each taxon; for example, a list of Vicieae species indexed by toxicity, with the presence or absence of specific chemicals given for each of the species in the list. The tailor-made services will be produced in liaison with members of the Study Group. They will be asked to provide lists of taxa and data in which they are particularly interested, and to suggest the type of product that may be useful to them. Products the user may request include phenetic diversity diagrams (partway between classification and ordination diagrams), phylogenetic classifications, printed or interactive identification aids, or printed descriptions.

So far we have had a wide variety of requests for information in the database, particularly for chemical information. For example, a plant geneticist has requested information about the presence of anthocyanins in closely related species of *Lathyrus*, and a chemotaxonomist in Turkey has asked for the distribution of non-protein amino acids and phytoalexins within the sections of *Vicia* and *Lathyrus*. An example of a request for non-chemical information is provided by a Portuguese member of the Study Group who requested a list of species reported from Portugal and morphological descriptions of these species in English and in Portuguese.

IMMINENT EXPERIMENTS

Specific aspects of our services and products are being monitored. For example, we will record the extent to which the tailor-made service is used by members of the Study Group, the subsets of data and species requested and the demand for the selection of classifications, tabulations, keys or diagrammatic representations of variation patterns available. We are also monitoring the time taken to produce a response to each request, and examining which of the various formats available for presenting the products (for example, cheap pamphlets or more permanent publications) are feasible to produce and satisfy users requirements. We have produced our first two publications (Allkin *et al.*, 1983a,b), lists of the species and subspecies in the Vicieae, with and without synonymies, and the next publication,

CONFOR-generated descriptions of species in *Vicia* section Faba, is in preparation.

CONCLUSIONS

(1) We have shown that it is possible to create and operate successfully a multipurpose monographic database, and to provide novel taxonomic services of use to the taxonomic community.

(2) We have sensed that there is a demand among the scientific community for efficient, rapid, up-to-date services not presently supplied by taxonomists, but we await the evaluation of the Project by the Study Group in September 1983 for further detail.

(3) The existing software, although adequate for our experimental project, is not suitable for general use; there is a need for an improved system which takes into account the needs of taxonomy and, specifically, the need to handle descriptive data adequately.

(4) We stand by our original hypothesis that it will prove worthwhile for the taxonomic profession to set up International Information Centres backed by monographic databases for the dozen or so groups of plants which are of great use to humans.

ACKNOWLEDGEMENTS

The authors are grateful to Dr M. J. Dallwitz for commenting on this paper, and to the SERC who funded the Vicieae Database Project with grants GR/A/78188 and GR/C10391.

REFERENCES

Abbott, L. A. (1976). "EXIR user's manual". University of Colorado, Boulder.

Allkin, R., Macfarlane, T. D., White, R. J., Bisby, F. A. and Adey, M. E. (1983a). "List of species and subspecies in the Vicieae". [Vicieae Database Project, Publication No. 1.] University of Southampton, Southampton.

Allkin, R., Macfarlane, T. D., White, R. J., Bisby, F. A. and Adey, M. E. (1983b). "Synonymised list of species and subspecies in the Vicieae". [Vicieae Database Project, Publication No. 2.] University of Southampton, Southampton.

Bisby, F. A. (1984). Automated Taxonomic Information Systems. *In* "Current Concepts in Plant Taxonomy" (V. H. Heywood and D. M. Moore, eds), pp. 301–322. Academic Press, London and Orlando.

Bisby, F. A., White, R. J., Macfarlane, T. D. and Babac, M. T. (1983). The Vicieae Database Project: experimental uses of a monographic taxonomic database for species of vetch and pea. *In* "Numerical Taxonomy" (J. Felsenstein, ed.), pp. 625–629. [NATO ASI Series G1]. Springer, Berlin.

Dallwitz, M. J. (1980). "Users Guide to the DELTA system. A general System for Coding Taxonomic Descriptions". [CSIRO Division of Entomology Report No. 13.] Division of Entomology, Commonwealth Scientific and Industrial Research Organisation, Australia.

Fleming, H. S. and Appan, S. G. (1972). "Cluster Analysis using Graph Theory". Gulf Universities Research Consortium, Bay St. Louis, Mississippi.

Kupicha, F. K. (1974). "Taxonomic studies in the tribe Vicieae (Leguminosae)". Ph.D Thesis, University of Edinburgh, Edinburgh.

Kupicha, F. K. (1976). The infra-generic structure of *Vicia. Notes R. b. Gd. Edinb.* **34**, 287–326.

Kupicha, F. K. (1982). The infra-generic structure of *Lathyrus. Notes R. b. Gd. Edinb.* (In press).

Pankhurst, R. J. (1976). "Key constructing program (version 2, mk. 4)". Internal report, British Museum (Natural History), London.

Tutin, T. G., Heywood, V. H., Burges, N. A., Moore, D. M., Valentine, D. H., Walters, S. M. and Webb, D. A. (1964–1980). "Flora Europaea" 5 vols. Cambridge University Press, Cambridge.

Watson, L. and Dallwitz, M. J. (1981). An automated data bank for grass genera. *Taxon* **30**, 424–429.

16 The Use of a Descriptive Database as an Aid to Assessing the Distinctness of Pea Cultivars *(Pisum sativum L.)*

P. J. WINFIELD and F. N. GREEN

Agricultural Scientific Services, Department of Agriculture and Fisheries for Scotland, East Craigs, Edinburgh EH12 8NJ, UK

Abstract: A database of all pea cultivars in the EEC Common Catalogue and United Kingdom National List is described. Its role as an aid to assessing Distinctness is discussed and its limitations and potential advantages are explored. The importance of using genetic data for devising characters and character states is emphasized.

INTRODUCTION

All pea cultivars marketed in the United Kingdom or European Community (EEC) must be registered by law on the United Kingdom National List (HMSO, 1964) or the EEC Common Catalogue (EEC, 1970). The Common Catalogue is a compendium of the National Lists of all countries belonging to the EEC. Once on any of the National Lists, a cultivar can be added to the Common Catalogue and marketed freely within the European Community. A cultivar must satisfy at least three conditions before registration, or before it may be awarded Plant Breeder's Rights (PBR):

(1) It must be Distinct from all listed cultivars where distinctness means "a clear, consistent difference in at least one morphological or physiological character in two similar growing periods".

Systematics Association Special Volume No. 26, "Databases in Systematics", edited by R. Allkin and F. A. Bisby, 1984, pp. 189–200. Academic Press, London and Orlando.

DATABASES IN SYSTEMATICS
ISBN 0 12 053040 6

(2) It must be Uniform, i.e. it must be sufficiently homogeneous, bearing in mind its method of reproduction.
(3) It must be Stable, i.e. the cultivar must maintain its characteristics from generation to generation.

In addition, peas on the Agricultural List must be shown to have value for cultivation and use.

Approximately 600 cultivars are registered on the lists. The number of candidates entered for registration, or Plant Breeder's Rights, varies between 20 and 50 each year. There are between 50 and 80 additions and deletions from the Common Catalogue each year.

In the United Kingdom the Distinctness testing of pea cultivars is carried out in two stages. First, information provided by the breeder, and some information gained from growing the cultivar, is used to classify it into a group of most similar cultivars – known as controls. If the candidate is classified into a group containing no other cultivars it is considered Distinct and can be registered. Since the behaviour of characters used to classify cultivars is consistent over site and season, it is assumed that the cultivar is Stable if the criterion for Uniformity is met.

If the candidate is not Distinct after the first stage it is grown in two comparative field trials. A randomized block design with two replications is used. Differences between candidates and controls are then measured or scored and analysed statistically, or noted as observations. A statistical criterion or a judgement based on the known behaviour of a character is then used to decide whether the candidate is Distinct.

Once accepted as Distinct, Uniform and Stable, a description of the candidate cultivar is published in the United Kingdom Plant Varieties and Seeds Gazette and its name is added to the National List. The cultivar name may subsequently be added to the Common Catalogue.

Testing cultivars for Distinctness is expensive in time and labour, particularly where large comparative trials must be grown. Any method which could be used to reduce the size of the comparative trials would reduce their cost. It would also reduce the time taken to assess Distinctness. An obvious way to do this is to classify the cultivars.

Establishing Distinctness is not simply a process of classifying cultivars. Many agronomic characters, e.g. days to flowering and stem height, vary over site and season. Many are also continuously distributed throughout the crop. These characters are useful for assessing Distinctness when a statistical criterion is applied to them within the context of an experimental design, but they are not necessarily useful for classifying cultivars because their variance cannot be adequately described.

A distinction therefore has to be made between characters which are useful for classification purposes and those which are valuable for asssessing Distinctness. There are likely to be far more characters suitable for the latter than the former. One problem which must be faced when using classification as an aid to Distinctness testing is in deciding which characters may be used in these classifications and which should be part of the experimental procedures. Since relatively few characters are used for classifying cultivars, large numbers of cultivars remain unclassified. The classsifications may be said to be only partially resolved.

The degree to which pea cultivars can be classified is illustrated in Table 1. The classification that has been produced is artificial; the purpose is merely to generate the smallest groups possible. The characters shown in Table 2 have been chosen because they are considered to be stable over sites and seasons. Most of these characters have been used in previous classifications of peas. As there are two character states for each character in Table 2, this would theoretically allow 8192 groups. In practice, most of these groups do not contain cultivars. Using 11 of the 13 characters we have produced 31 groups (Table 1). Although there are 19 groups with fewer than 10 cultivars, the six largest groups contain between 36 and 134 cultivars. Most candidates entered for test fall into one of these large groups.

A further problem is a practical one. To be used effectively as a tool for assessing Distinctness such classifications must be both easy to revise and generated relatively frequently. This was difficult to do manually, which meant that the classifications were not as useful as they could have been. Consequently, the comparative trials tended to be very large.

To overcome these problems it was decided to create a database which would, it was hoped, help to reduce the size and cost of the comparative trials and provide some form of information service for breeders and genetic conservationists.

The creation of the database in turn stimulated questions concerning what criteria, if any, could be used in selecting characters. Finally, the poor resolution of the classification prompted consider-

Table 1. Numbers of pea cultivars in 31 major groups generated using 11 characters. Character numbers refer to those used in Table 2. Each character state is represented by its first letter. A dash indicates that the character was not used to generate the group.

Character 10	9	2	1	6	11	4	7	12	13	5	*Listed Cultivars*
sativum											
s	b	y	a	p	n	p	p	a	a	n	25
s	b	y	a	p	n	d	p	a	a	n	2
s	b	g	a	p	n	p	p	a	a	n	74
s	b	g	a	p	n	d	p	a	a	n	20
s	p	y	a	p	n	p	p	a	a	n	14
s	p	y	a	p	n	d	p	a	a	n	2
s	p	g	a	p	n	p	p	a	a	n	47
s	p	g	a	p	n	d	p	a	a	n	17
c	b	y	a	p	n	p	p	a	a	n	3
c	b	y	a	p	n	d	p	a	a	n	4
c	b	g	a	p	n	p	p	a	a	n	55
c	b	g	a	p	n	d	p	a	a	n	134
c	b	y	a	p	n	p	p	a	a	n	6
c	p	y	a	p	n	d	p	a	a	n	10
c	p	g	a	p	n	p	p	a	a	n	36
c	p	g	a	p	n	d	p	a	a	n	66

Character 10	9	2	1	6	11	4	7	12	13	5	*Listed Cultivars*
saccharatum											
s	—	y	a	—	—	—	a	—	—	—	20
s	—	g	a	—	—	—	a	—	—	—	1
c	—	y	a	—	—	—	a	—	—	—	4
c	—	g	a	—	—	—	a	—	—	—	7
—	—	—	p	—	—	—	a	—	—	—	7
arvense											
—	—	—	p	—	—	—	p	p	p	n	1
—	—	—	p	—	—	—	p	p	a	b	6
—	—	—	p	—	—	—	p	p	a	n	12
—	—	—	p	—	—	—	p	a	p	b	3
—	—	—	p	—	—	—	p	a	p	n	9
—	—	—	p	—	—	—	p	a	a	b	1
—	—	—	p	—	—	—	p	a	a	n	6
leafless and semi-leafless											
—	—	—	—	a	n	—	—	—	—	—	8
—	—	—	—	a	r	—	—	—	—	—	1
—	—	—	—	p	r	—	—	—	—	—	2

ration of how it might be improved and how the bias of testing for Distinctness could be shifted from a trial and experimental orientated scheme to one which relied more on the use of classifications and databases.

Table 2. Thirteen major characters used to classify pea cultivars. The expression of these characters is not markedly influenced by environmental conditions.

	Character	*Character States*	*Gene(s) Responsible*
1	Anthocyanin pigmentation	present/absent	A
2	Cotyledon colour	yellow/green	i
3	Foliage marbling	present/absent	fl
4	Fresh seed colour	pale/dark green	Pa, vim
5	Hilum	black/not black	Pl
6	Leaves	present/absent	af
7	Pod parchment	present/absent	p, v
8	Pod pigment	present/absent	Pu, Pur
9	Pod tip shape	blunt/pointed	bt
10	Starch grains	simple/compound	r, Rb
11	Stipules	normal/reduced	St
12	Testa marbling	present/absent	M
13	Violet spotting on testa	present/absent	F, Fs

CHOICE OF CHARACTERS

When designing a database decisions must be taken about which characters and character states are to be included. These decisions are critical. The resolution of a classification depends on the discriminating power of the characters. Simplifying, we can say that those characters which will be most useful and easiest to record, will have character states which are clearly defined and easily recognized. Preferably, such characters should be relatively stable over sites and seasons, although they may still be of value if their behaviour is predictable. An example of such a character is the violet pigmentation sometimes observed over the entire surface of the testa of coloured flowered peas. This character is only expressed when the seeds are subjected to intense heat and light. Thus it is only seen in particularly hot summers. Although the expression of this character is variable over site and season, it is predictable. It could therefore be useful as a classification character provided a test for the presence of pigment could be devised.

The most useful source of such characters in some crops is the information which has been collected from genetic studies. Simple Mendelian inheritance is concerned with characters which are clearly defined and easy to see, and the study of inheritance provides a critical method for examining the definition of characters and the relationship between character states.

Much genetic information has been collected for peas. Nearly 2000 mutants of *Pisum* have been reported and, of the roughly 500 known genes, approximately 170 are mapped and their mode of inheritance understood (Wellensiek, 1925; Blixt, 1972, 1974). This information has been codified (Blixt, pers. comm.). In addition to a standard list of gene symbols, phenotypic descriptions are available for each of the known genotypes (Blixt, 1977). Lines in which the genes were first isolated (type lines) and lines which most clearly show the expression of the gene (representative lines) are also maintained.

THE DATABASE

1. Scope

The database contains information on approximately 800 pea cultivars including all those listed on the United Kingdom National List and the European Common Catalogue, cultivars which are in test and a number which are no longer in commerce. The database has 27 descriptors (Table 3). Three of them, cultivar, supplier and status, provide simple administrative information concerning the cultivar. The remaining 23 descriptors describe the morphology and agronomic characteristics of the cultivar. Thirteen of these are used to classify cultivars, the rest are commercially important characters which are useful for discriminating cultivars.

2. Software and Hardware

The Executive and Information Retrieval Package (EXIR) Version 3.2 originally written at the University of Colorado, USA (Brill and Estabrook, Chapter 5, this volume), is used to maintain the database. EXIR is mounted on an IBM computer of the Scottish Office Computer Services. The package is accessed in batch mode using Remote Job Entry via a terminal situated at Agricultural Scientific Services.

3. Style of Input

The database is interrogated in two ways when used to make decisions about which cultivars are to be grown in trial. In the first instance, partial classifications can be produced by submitting a file containing a standard set of interrogations to EXIR. These classifications are artificial; the objective being to produce the smallest possible size of group. These interrogations use all combinations of classifying characters which produce groups containing cultivars. The partial classification lists all cultivars in each of these groups along with their agronomic characters. The groups are also sorted in

Table 3. Descriptors used on the database of pea cultivars.

Descriptor	*Type*
Cultivar	
Supplier	
Status	
Leaves	score
Leaf colour	score
Stipules	score
Stipule marbling	score
First fertile node (FFN)	metric
Flower pigment	score
Flower number at FFN	metric
Peduncle length	metric
Days to flowering	metric
Days to 80% flowering	metric
Pod parchment	score
Pod tip shape	score
Pod colour	score
Ovules plus seeds	metric
Fresh seed colour	score
Fresh seed size	metric
Starch grains	score
Cotyledon colour	score
Hilum colour	score
Violet Spotting on testa	score
Testa marbling	score
100 seed weight	metric
Resistance to Pea Wilt (Race 1)	score

different character orders. The second method of selecting controls is to construct a profile of the candidate and generate an interrogation to list all similar cultivars. To speed up this process a menu-driven program has been written which builds EXIR interrogation requests.

4. Output

The following standard outputs are produced: partial classifications, lists of all cultivars on the Common Catalogue and UK National List and cultivar descriptions. The latter cannot be produced directly by EXIR but a general purpose program has been written which reads output produced by EXIR and creates a simple list of all the descriptors and descriptor states for the desired cultivar or cultivars. The partial classifications and listings are reprinted whenever the database has been revised which is usually several times a year. Cultivar descriptions are printed as and when they are required. In addition to these standard outputs various ad hoc interrogations are produced in response to queries by breeders and other test authorities.

DISCUSSION

Using EXIR we have succeeded in our primary purpose of providing a rapid method of producing partial classifications. These outputs can be revised easily and printed frequently. This is a considerable advance over older methods of producing such information. The use of these classifications has had a considerable impact on our comparative trials, almost halving their size.

The ease with which a variety of interrogation requests can be made and the rapidity with which such requests may be answered, also makes it possible to provide an information service to plant breeders. The database can be used to identify unexploited niches for cultivars and to determine, in advance of submitting a cultivar for test, what cultivars are likely to be close to a candidate. This information can be provided either in the form of check-lists or as printouts of specific interrogation requests from breeders.

The partial classifications that we are able to produce have a lower resolution than those that have been previously published. Part of the reason for this is that earlier classifications are regional or national in scope and, consequently, include fewer cultivars; but the main reason is that they use more characters. However, these characters include those which are simply inherited and have high heritabilities and those which are continuously distributed and variable, e.g. stem height (Cousin, 1974; Fourmont, 1956; Sneddon and Squibbs, 1958) and pod size (Wade, 1943). This means that the classifications cannot be used reliably and independently of a comparative trial (Schoemaker and Delwische, 1934).

We have made a more rigid distinction between these two types of character. When defining the database we have mainly chosen those characters which are monogenically or oligogenically inherited. Those characters which do not fall into this category are stored as a mean and standard deviation and are used to determine Distinctness within the context of a comparative trial. They are not used for classifying cultivars.

There is still a great deal of potential for exploiting genetic data as a source of characters for classifying peas. This data can be useful in three ways.

First, it will suggest new characters. Although genes with easily detected or clearly defined effects only constitute a small part of the total heritable variation (Marx, 1977), between 30 and 40 of the 170 mapped genes may be useable in this respect; considerably more than the 13 used presently. In a recent investigation six previously unused seed characters were recorded for 54 coloured flowered cultivars. Preliminary results suggest that 33 can be identified using seed characters alone (Winfield and Green, unpublished). Previously only five cultivars could be identified using seed characters.

Secondly, the genetic data may help to clarify the treatment of characters. Stem height, for example, has been used as a primary classifying character. Cousin (1974) arbitrarily divided the crop into two classes: dwarf/semi-dwarf and tall. Stem height is continuously distributed and variable over years, which makes the character difficult to use. Instead of treating stem height as a simple metric character it may be more profitable to examine its genetic components. Marx (1977) discusses the interaction of three genes which determine internode length, a component of stem height, and was able to recognize four phenotypes. Cousin (1974) lists 10 genes responsible for the character. One major gene, *le*, produces a characteristic zig-zag pattern in the stem. This was used by Schoemaker and Delwische (1934) for classifying peas. The genetic data suggests that some of the components of stem height would be more appropriate for classification than stem height itself.

Thirdly, the relationships between the character states may also be clarified. An example of this is the character pod parchment, which is absent in edible podded peas. The guidelines produced by the International Union for the Protection of New Varieties of Plants (UPOV) for Distinctness, Uniformity and Stability tests list two states for pod parchment, present and absent (UPOV TG/7/4). The genetic evidence suggest two genes, *p* and *v* (Mendel, 1866; White, 1917), are responsible for the removal of parchment from the pod wall. Four, rather than two, phenotypes can be distinguished (Table 4).

Table 4. Genotypes and phenotypes for the genes *p* and *v*.

Genotype	*Phenotype*
P V	normal parchmented pod
p V	strip of sclerenchyma
P v	inner membrane reduced to patches of sclerenchyma
p v	entirely without parchment

To make full use of this genetic data we intend to define characters and character states by selecting from the literature all those major genes which are clearly expressed and easy to see, and which have not yet been included in our character lists. We then intend to grow representative or type lines of each phenotype and, as far as is possible, characterize the cultivars in the reference collection. The inclusion of these new characters will improve the resolution of our classifications which will bias testing towards a database oriented scheme.

Using genetic data to define characters and character states will make this database of greater value to breeders and to genetic conservationists interested in particular genotypes. The model used in defining this database conforms to that proposed by Blixt *et al.* (1982) for a gene databank. As more characters are added to the database EXIR will become less useful for classifying cultivars although it will retain its importance for data management. EXIR is a data management package and cannot be used to generate classifications automatically.

Where cultivars are classifiable it will be possible to use programs for automatic key generation and on-line identification. Indeed, such programs, e.g. those written by Pankhurst (1978), have already been used successfully to generate keys and descriptions of pea cultivars with coloured flowers (Winfield and Green, unpublished) using only seed characters. We foresee that in the future these and other programs (e.g. those used for Cluster analysis) will be integrated into a computing environment for manipulating and analysing data on pea cultivars.

CONCLUSIONS

As far as we are aware this is the first database to be used as an aid to Distinctness testing. We feel that databases have an important role to play in assessing Distinctness in some crops. They provide a means of managing data efficiently and may also be used in conjunction with other programs to generate keys and classifications automatically. Such databases can also be used to provide information for breeders and genetic conservationists.

When creating such databases particular care must be taken in the selection of characters. Genetic data are useful for defining characters and character states. Use of this data will ensure a greater conformity between the characters used by genetists and those used for classifying cultivars. Such data should also produce reliable classifications with a higher resolution and which are independent of site and season.

REFERENCES

Blixt, S. (1972). Mutation Genetics in *Pisum. Agri. Hort. Genet.* **30**, 1–293.

Blixt, S. (1974). The Pea. *In* "Handbook of Genetics" (R. C. King, ed.) Vol. II, pp. 181–221. Plenum Press, New York and London.

Blixt, S. (1977). Gene symbols of *Pisum. Pisum Newsletter* **9** (suppl.).

Blixt, S., Yndgaard, F. and Kjellqvist, E. (1982). "A structure Model for Gene Bank Data". [Nordic Symposium in Applied Statistics and Data Processing.] NEUCC, Danish Technical University, Denmark.

Cousin, R. (1974). "Le Pois. Étude Génétique, des Caractères, Classification, Caractéristiques Varietales portant sur les Variétés inscrites au Catalogue Officiel Français". Institut National de la Recherche Agronomique, Paris.

EEC (1970). "Council Directive". Nos. 70/457/EEC and 70/458/EEC, Luxembourg.

Fourmont, R. (1956). "Les Variétés de Pois Cultives en France". Institut National de la Recherche Agronomique, Paris.

HMSO (1964). "Plant Varieties and Seeds Act 1964". Her Majesty's Stationery Office, London.

Marx, G. A. (1977). Classification, Genetics and Breeding. *In* "The Physiology of the Garden Pea" (J. F. Sutcliffe and J. S. Pate, eds), pp. 21–44. Academic Press, London and Orlando.

Mendel, G. (1866). Versuche uber Pflanzenhybriden. *Verh. naturf. Ver. Brunn.*, 4, 3–47. [*In* Ostwald's Klassiker der exacten Wissenshaften, **121**, 1940.]
Pankhurst, R. J. (1978). "Biological Identification". Edward Arnold, London.
Shoemaker, D. N. and Delwische, E. J. (1934). Descriptions of Principal American Varieties of Garden Peas. *Misc. Publ. USDA* **170**.
Sneddon, J. L. and Squibbs, F. L. (1958). Classification of Garden Pea Varieties. *J. Nat. Inst. Agric. Bot.* 8, 378–422.
UPOV (1981). "Guidelines for Distinctness, Homogeneity and Stability: TG/7/4 Peas". International Union for the Protection of New Varieties of Plants, Geneva.
Wade, B. L. (1943). A key to pea varieties. *USDA Circular* **676**.
Wellensiek, S. J. (1925). Genetic monograph on *Pisum*. *Biblphia. Genet.* **2**, 343–476.
White, O. E. (1917). Studies of inheritance in *Pisum*. II. The present state of knowledge of heredity and variation in Peas. *Proc. Am. phil. Soc.* **56**, 487–588.

17 | A Chemical Database for the Leguminosae

*B. V. CHARLWOOD, G. S. MORRIS** *and M. J. GRENHAM*

Department of Plant Sciences, King's College London, London SE24 9JF, UK

Abstract: Legumes are a very rich source of amino acids since they contain not only the 20 or so protein amino acids but also some 250 analogues, homologues or derivatives that are not normally incorporated into proteins. It is generally accepted that a study of this group of compounds can provide valuable information concerning the relationship within and between different genera, tribes and families of plants. Much of the information at present available concerning the amino acid content of species of the Leguminosae is of a qualitative nature. We have analysed the seeds of legumes in a strictly quantitative manner in order to establish if such data could be of greater value in studying taxonomic relationships within the Leguminosae. The paper describes the construction of a quantitative chemical database involving over 120 species from 29 genera in the Mimosoideae, and some 170 species and varieties of the single genus *Indigofera*. The application of a 6502 based microcomputer to the problem of pattern recognition within the database is discussed and our conclusions concerning the potential of this method presented.

INTRODUCTION

This paper describes the use of a chemical database constructed from the results of the quantitative analysis of the free protein and non-protein amino-acid pools in seeds of various species of the Leguminosae.

---

* Present address: Clinical Sciences Laboratories, Guy's Hospital, London Bridge, London SE1 RT, UK.

Systematics Association Special Volume No. 26, "Databases in Systematics", edited by R. Allkin and F. A. Bisby, 1984, pp. 201–208. Academic Press, London and Orlando.

DATABASES IN SYSTEMATICS
ISBN 0 12 053040 6

All organisms utilize some 20 amino acids and derivatives as building blocks in protein production, but some plants and microorganisms have the ability to manufacture over 250 further compounds which may be formally considered amino acids. These compounds are not of primary importance to the individual organism in the same way that the protein components are, nor are they universally distributed throughout the plant kingdom. Such unusual amino acids tend to be analogues, homologues or bizarre derivatives of their protein counterparts, and the term "non-protein" amino acid has been coined to describe what are essentially secondary compounds. Aspects of the distribution, biosynthesis and ecological significance of the non-protein amino acids have been reviewed by Bell (1980) and Fowden (1981), and many examples are available in which a knowledge of non-protein amino acid content has been of use in the elucidation of relationships within and between different species, genera and tribes.

Many of the taxonomic studies involving non-protein amino acids that have been reported, have involved qualitative ("presence" or "absence") analysis. The reason for this fairly simple approach is that non-protein amino acids may easily be detected after paper chromatography or high-voltage paper electrophoresis by the application of ninhydrin reagent. In many cases the colours of the spots produced with ninhydrin may themselves provide a clue as to the identity of the amino acid involved. Furthermore, in certain cases, specific chemical tests have been devised that will produce coloured products with just one or a small group of amino acids, and such tests eliminate the need to carry out the preliminary chromatographic separation. Unfortunately, a reliance upon qualitative analysis implies a reliance upon the sensitivity of the detection system to determine the threshold concentration at which any particular amino acid is claimed to be present. This threshold level will be different for each component present, and will also vary for the same compound with each different set of detection conditions.

In view of the possible disadvantages inherent in the use of qualitative data for taxonomic purposes, we have analysed extracts of seeds of species of the Leguminosae by quantitative techniques. Completely automated methods of analysis of free amino-acid pools present in plant extracts have been developed in our laboratories (Charlwood and Bell, 1977), together with a computer-based data-handling system (Charlwood and Bell, 1980) that acts as a datalogger, an on-line integrator and for mass-storage. This system has been further extended to include database file handling and interrogation.

CONSTRUCTION OF THE DATABASE

The computer configuration finally adopted for automatic analysis and data storage is built around a Commodore Business Machines' PET Model 8032 (a 6502-based microcomputer with an 80-column visual display and 32 Kbytes of user RAM). The computer is connected to an LKB Model 4400 automatic amino acid analyser via an analogue-to-digital converter for data acquisition, and to a dual drive floppy disc (CBM Model 8050) for data logging and database management. The double-density disc drive has a memory capacity of over 1 Mbyte and, hence, the total package is fairly powerful in terms of file handling. The final amino-acid profile for each specimen analysed is coded and can be stored, together with associated protocol, using less than 200 bytes; ten such profiles are stored in each indexed file. Access time for any particular profile is less than 3 s and so the file interrogation is reasonably fast for such a small system. Simple BASIC language programs have been written for all the standard forms of data interrogation, and these are presented as an initial menu from which individual selections can be acquired from disc. With this technique the user has the impression that the computer memory is of the same capacity as that of the floppy disc, but the penalty to be paid is in increased access time.

We have analysed quantitatively some 120 species representing 29 genera of the subfamily Mimosoideae, as well as over 170 species and varieties of the single genus *Indigofera* (subfamily Papilionoideae) at the present time. This paper reports some of the results obtained from our investigations concerning the genus *Indigofera*.

NON-PROTEIN AMINO ACIDS IN THE GENUS *INDIGOFERA*

There are two important non-protein amino acids in the *Indigofera* namely canavanine (I) and indospicine (II), both of which are analogues of the protein amino acid arginine (III). Canavanine was one of the first arginine analogues to be found in plants, and its toxicity in microorganisms and animals has been widely studied. It appears that whilst canavanine is fairly toxic to some insects, it is relatively harmless to animals at concentrations that might reasonably be expected to be attained in the field. On the other hand indospicine, which was first isolated from *Indigofera spicata* (Hegarty and Pound, 1968), is very toxic to a wide range of animals and is both an hepatotoxin and a potent and highly specific teratogen.

$$\text{(I)}\quad H_2N\overset{\overset{NH}{\|}}{C}NHOCH_2CH_2\overset{\overset{NH_2}{|}}{C}HCOOH$$

$$\text{(II)}\quad H_2N\overset{\overset{NH}{\|}}{C}CH_2CH_2CH_2CH_2\overset{\overset{NH_2}{|}}{C}HCOOH$$

$$\text{(III)}\quad H_2N\overset{\overset{NH}{\|}}{C}NHCH_2CH_2CH_2\overset{\overset{NH_2}{|}}{C}HCOOH$$

In collaboration with the Commonwealth Scientific and Industrial Research Organisation in Brisbane, Australia, we have analysed nearly two-hundred samples of seed material from various species of *Indigofera* that have been collected from sites throughout Australia. In view of the toxic nature of indospicine, our primary interest was to study the distribution of this compound within the genus with a view to selecting non-toxic species as possible forage crop plants.

Table 1 shows the concentrations of indospicine (expressed in micromoles of amino acid per gram of seed) found in the seed samples in our collection. The table shows that only 17 seed samples contain even trace quantities of indospicine, and although the compound is often found in *I. spicata*, less than one-quarter of the seed samples of this species that we have analysed actually contained detectable amounts. It is clear that the occurrence of indospicine is very variable both qualitatively and quantitatively throughout the genus, and even within each species, and it is likely that any information concerning the content of this amino acid would be of little taxonomic use.

Examination of the amino acid profile database has also indicated the presence of three further non-protein amino acids, the structures of which have yet to be confirmed. But, as with indospicine, it would seem that these alone will not be of much help in determining species identity within the genus.

In contrast with the situation for indospicine and the other non-protein amino acids, the concentrations of canavanine in seeds of *Indigofera* species appear to be very constant. We have been able to demonstrate that the pool size of canavanine within the seed is not

Table 1. Indospicine content of seeds of *Indigofera* species.

	Seed Acquisition[a] *Number*	*Indospicine*[b] *Content*
Indigofera circinella (2)[c]	51424	trace
I. heterotricha (1)	71069	0.43
I. hirsuta (32)	29073	trace
I. hirsuta	33430	trace
I. hirsuta	79610	0.51
I. hirsuta	CQ147	0.21
I. lespedezioides (3)	34167	0.64
I. spicata (35)	16110	trace
I. spicata	24205	trace
I. spicata	28827	0.86
I. spicata	29295	0.43
I. spicata	29485	trace
I. spicata	33818	trace
I. spicata	37384	0.81
I. spicata	37415	trace
I. spicata	38211	0.35
I. trita (15)	65478	1.70

[a] Voucher specimens held by CSIRO, Brisbane, Australia.
[b] Amino acid content expressed in micromoles per gram of seed.
[c] Number of acquisitions of this species analysed.

dependent upon the growing environment of the plant in the same way that proline or glutamine pools appear to be. Figure 1 shows a histogram representing the average concentrations of canavanine in seeds of all of the *Indigofera* species analysed thus far.

It appears that all species of *Indigofera* contain canavanine to some extent. Bell and other workers (Bell *et al.*, 1978; Fowden *et al.*, 1979) have shown that canavanine is confined to the subfamily Papilionoideae, and have constructed a possible phylogenetic tree, based on canavanine content, indicating the relationships between the various tribes within the Papilionoideae. It would be expected from this hypothesis that the genera of the Indigofereae, a small but advanced tribe of the subfamily, would be particularly rich in canavanine, and our present results concerning *Indigofera* are in complete agreement with this.

The amounts of canavanine appear to be very specific for each species, and are quite conservative within each species. The range of

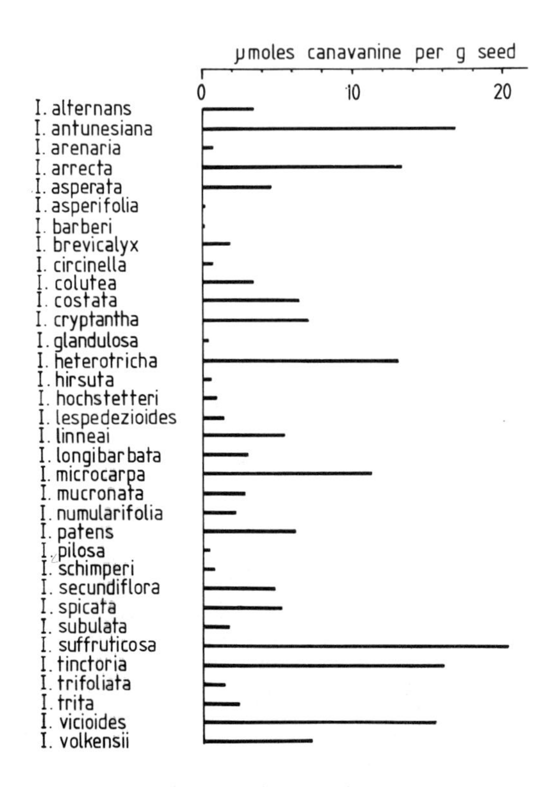

Fig. 1. Canavanine content of seeds of *Indigofera* species.

canavanine conentration within the genus as a whole is very large. Species such as *I. antunesiana* and *I. suffruticosa* can contain up to 20 micromoles of canavanine per gram of seed, and this represents something in the region of 90% of the total free amino-acid pool within the seed. At the other extreme, *I. asperifolia* and *I. barberi* contain only trace quantities of the non-protein amino acid.

PROFILES OF THE FREE AMINO ACID POOLS
IN THE GENUS *INDIGOFERA*

At a preliminary stage in this quantitative study of the free amino-acid pools in seeds it appeared that the overall profile of the pool (including both protein and non-protein amino acids) might, in itself, contain unique information at the specific or generic level. This turned out not to be the case, at least at the superficial level. However, further interrogation of the database has revealed that there are a number of close associations, in quantitative terms, between various of the protein and non-protein amino acids within each species. Many of these associations are unique to a particular species and may be useful for species recognition.

Some particularly strong associations are shown in Fig. 2 which

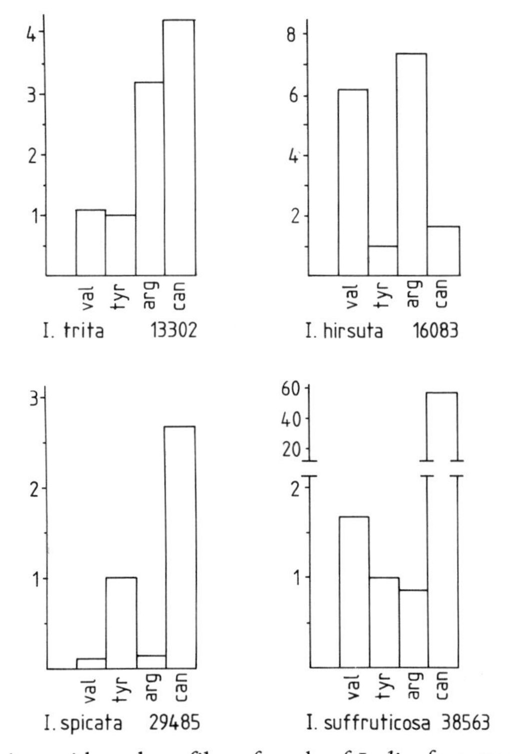

Fig. 2. Partial amino-acid pool profiles of seeds of *Indigofera* species.
NB. The height of a bar represents the concentration of the corresponding amino acid relative to tyrosine.

depicts part of the profile of the amino-acid pools of various species of *Indigofera*. Each bar of each chart represents the pool concentration of a particular amino acid expressed as a ratio value with respect to tyrosine. It is clear that the "part profile" for each species is totally different from that of every other species shown.

Similar "part profiles" have been constructed for all *Indigofera* species examined in this study. In order to demonstrate that the "part profile" patterns are conservative within each species, the Canberra metric and the Bray-Curtis coefficients of dissimilarity have been calculated for these patterns for all seed acquisitions taken in pairs. The results show that within species both coefficients are less than 0.1, whilst across species values lie in the range 0.4 to 0.9.

It is likely that some of these protein/non-protein amino acid associations will be of value in the determination of biosynthetic routes or control, and further work is continuing in this area.

ACKNOWLEDGEMENTS

We would like to thank Joyce E. Potter for invaluable technical assistance, and LKB Biochrom Ltd, Science Park, Cambridge, UK for the kind loan of a Model 4400 automatic amino-acid analyser.

REFERENCES

Bell, E. A. (1980). Non-protein amino acids in plants. *In* "Encyclopedia of Plant Physiology" (E. A. Bell and B. V. Charlwood, eds), Vol. 8, pp. 403–432. Springer-Verlag, Berlin.

Bell, E. A., Lackey, J. A. and Polhill, R. M. (1978). Systematic significance of canavanine in the Papilionoideae (Faboideae). *Biochem. Syst. Ecol.* **6**, 201–212.

Charlwood, B. V. and Bell, E. A. (1977). Qualitative and quantitative analysis of common and uncommon amino acids in plant extracts. *J. Chromat.* **135**, 377–384.

Charlwood, B. V. and Bell, E. A. (1980). Automatic amino acid analysis: Data handling. *In* "Chemosystematics: Principles and Practice" (F. A. Bisby, J. G. Vaughan and C. A. Wright, eds.), pp. 91–102. Academic Press, London and Orlando.

Fowden, L., Lea, P. J. and Bell, E. A. (1979). The nonprotein amino acids of plants. *Adv. Enzymol.* **50**, 117–175.

Fowden, L. (1981). Nonprotein amino acids. *In* "The Biochemistry of Plants" (E. E. Conn. ed.), Vol. 7, pp. 215–247. Academic Press, Orlando and London.

Hegarty, M. P. and Pound, A. W. (1968). Indospicine, a new hepatotoxic amino acid from *Indigofera spicata*. *Nature. Lond.* **217**, 354–355.

18 | A Chemotaxonomic Database

M. T. BABAÇ

Biyoloji Bölümü, Fırat Üniversitesi, Elazığ, Turkey

and

F. A. BISBY

Biology Department, The University,
Southampton SO9 5NH, UK

Abstract: An experimental chemotaxonomic database has been created for occurrences of secondary substances judged of taxonomic value for 111 species of vetch and pea (Leguminosae tribe Vicieae). Data entries are arranged both separately by plant organs and as an amalgamated entry for whole plants. Substances entered include non-protein amino acids, flavonols, anthocyanidins and substances detected as phytoalexins. For each substance a data entry records whether it has been detected in a given organ of a given species, and provides references to the source of information. Much attention, including repeat analyses and cultivation work, has been given to overcoming the difficulties of data consistency, authentication of materials and biological variability. Data enquiries, tabulations and subset retrievals are carried out by the EXIR program. Of particular value is the ability to tabulate occurrences of selected substances for particular organs in selected species. Data subsets can also be retrieved using EXIR and passed on to other programs to perform taxonomic analyses such as cluster analysis.

INTRODUCTION

There has been a rapid accretion in the literature of records of the occurrences of chemicals in plants. The distribution in different species has been reported for a very wide range of substances and, in

Systematics Association Special Volume No. 26, "Databases in Systematics", edited by R. Allkin and F. A. Bisby, 1984, pp. 209–218. Academic Press, London and Orlando.

DATABASES IN SYSTEMATICS
ISBN 0 12 053040 6

some cases, these distributions cover a taxonomically meaningful sample of taxa. That these distributions may be of use in determining the classification or the phylogeny of organisms is now well established (Bisby *et al.*, 1980) and it is commonplace for systematists to use chemical evidence alongside other traditional types of evidence (Smith, 1976).

It may seem surprising then that so little has been done to provide collated compendia, either from which taxonomists may obtain data for taxonomic analysis or by which taxonomically arranged chemical data can be disseminated for use by other scientists. However, the few compendia that do exist, for instance, *Chemotaxonomy of Flowering Plants* (Gibbs, 1974) or *Chemotaxonomy of the Leguminosae* (Harborne *et al.*, 1971) illustrate difficulties which may explain this shortage. Records accrue at such a pace that a published compendium soon becomes out of date. Selecting and collating records for comparison presents many difficulties involving the authentication and nomenclature of both plants and substances.

In 1978, we started the design and implementation for a chemotaxonomic computer database for secondary substances detected in vetch and pea species (the tribe Vicieae, in the Leguminosae). Our purpose was to use this experimental database (i) to provide a compendium that could be kept up-to-date; (ii) to explore solutions to the difficulty in collating comparable records; and (iii) to experiment with retrieving data from the compendium to meet user's requirements or for taxonomists to analyse.

We chose the Vicieae primarily because of the chemical and chemotaxonomic interest of the group and because of the intention to link this activity with the Vicieae Database Project (Bisby *et al.*, 1983; Adey *et al.*, Chapter 15 this volume). Chemical interest lies in the floral pigments such as those of Sweet Peas, *Lathyrus odoratus*, toxins such as non-protein amino acids involved in lathyrism, phytoalexins of *Vicia faba* and in human blood-group specific lectins found in *Vicia graminea* and other species. The distributions of non-protein amino acids and leaf flavonoids and of phytoalexins have been used in discussion of the taxonomy of *Vicia* and *Lathyrus*.

THE DATABASE

After a series of experiments with different designs, we decided to organize data on chemical occurrences in a rectangular table for storage and retrieval by the EXIR 3.0 program (Abbott, 1976).

The rectangular table has several rows allocated to each species: one for each of several different organs and one a combined entry for all parts of the plant. Thus for *Vicia bithynica* there are rows for Flower, Leaf, Seed and another for the whole plant. There are 111 of these groups for the 111 species, plus some extra entries for subspecies. Thus *Vicia caroliniana* has a single group of entries whereas *V. cracca* has a group of entries for the whole species, and separate groups of entries for its four subspecies.

Each column of the rectangular table contains data for one descriptor or chemotaxonomic character. Eleven of these descriptors contain nomenclatural or classificatory data as in the descriptor Genus which contains the generic names appropriate for each entry, *Vicia*, *Lathyrus*, *Pisum*, or *Lens*. Six of the descriptors contain simple data of biological interest such as flower colour: Purple, Red, etc. The remaining 210 descriptors contain data for chemical substances.

For each substance there are two and, in some cases, three descriptors. The first is a record of whether the substance is recorded present, recorded not present, or unobserved for the organ or whole plants of the taxon in question. The second is a coded entry of the source of the record, usually a literature citation. The third, present only for certain substances, is a crude relative estimate of quantity. The three entries for the non-protein amino acid (NPAA) lathyrine (Lat) have the following descriptors and descriptor states:

(Presence)	Lat/NPAA: Unknown (i.e. unobserved), Not detected, Present.
(Quantitative)	Qnt Lat/NPAA: Unknown, Not detected, Trace, Low, Medium, High, Variable.
(Source)	Ref Lat/NPAA: any coded citation such as Be62 (Bell, 1962).

The present records are for 29 non-protein amino acids (descriptor names coded NPAA as in the example above), 2 protein amino acids (PAA), 2 cinnamic acids (CNCA), 2 flavonols (FLVL), 37 phenolics (PNLC), 1 leucoanthocyanidin (LeuAncn), 9 phytoalexins (PALEX), and 3 Anthocyanidins (ANCN). The "Unknown" entry is used when an organ has not been surveyed for a substance and the "Not detected" entry is used when it has been surveyed but the substance was not detected.

DATA COLLATION

If the distributions of substances are to be of value in chemosystematics then the accurate determination of plant materials and the reliable comparability of chemical records are of importance. Absolute precision could only be achieved by one person or team working on the identification of materials and surveying for many classes of substance. Collating records for the same taxa but for different classes of substances surveyed on different materials by different workers involves some compromise. One of our aims was to see whether a combination of careful scrutiny and selection of other workers' records, with some infilling and test of comparability by ourselves, would enable us to compile a database of reasonably high accuracy and useful comparability.

Reliable determination of plant materials is ensured for our own chemical records by germinating and growing all seed stocks used and preparing a voucher specimen which is carefully identified and placed in the herbarium (SPN). With records obtained from the literature or other research groups there is the more difficult task of evaluating the likelihood of the names being correct. This is a subjective and sometimes embarrassing exercise. We selected records where (i) the authors had taken their material to an acknowledged expert in the group for identification or obtained it from them, or (ii) had deposited voucher specimens. In general, authors who had surveyed many species and had made taxonomic conclusions tended to have taken care with identification. We omitted records from authors who reported for only one or two species in our group, reported the same species several times under well-known synonyms, or who acknowledged that they had depended on seed packet labels for identifications.

To achieve comparability between records for the same substance in different taxa, we selected records from authors who were acknowledged experts in the group of substances in question, who had completed extensive surveys of species, and who had given details of the techniques used. This policy sometimes favoured less detailed work of extensive comparative studies over detailed analyses of single species. Where this policy provided two or more comparative studies of the same substances then we performed our own analyses to check for comparability. We found in tests with sample species that we could reproduce results which agreed accurately with those of Bell

(1962) and Bell and Tirimanna (1965) for non-protein amino acids, and with Torck *et al.* (1971) and Gibbs (1974) for the Flavonols Quercetin and Kaempferol. As a result we were able to combine records from these sources with records from our own observations and so fill in many gaps in the database. Detailed analyses of single species provided a variety of problems: many provide data on the presence of substances in one species for which there are no data in other species; if entered in the database they waste space, as a new descriptor is recorded "unknown" for all remaining species; and most of these records relate only to the four major cultivated species in our group. The occurrence of a substance in one species is of chemotaxonomic importance only if the substance has been surveyed and found missing from other species.

The chemical records collated for the database were those judged of possible chemotaxonomic value within the Vicieae. They were for substances that had patterns of occurrence involving several of the species. Substances present in all species recorded or absent in all species recorded were omitted.

DATABASE APPLICATIONS

1. Data Retrieval

The EXIR program (see Brill and Estabrook, Chapter 5, this volume) gives on-line responses to requests for particular data items, for searches, or for catalogued data. For instance, a request for particular data items, say the presence of malvidin and the source of this information for *Vicia cracca* would yield the response "not recorded, Pe60", where reference to another file would yield the details of the reference to Pecket (1960). A search would be, say, for species with both purple flowers and malvidin present. The two logical parts to the retrieval command, the search or location part (which locates species with a stated combination of data entries), and the examination part (that specifies the data of interest once the species has been located), allow quite a variety of lists or catalogues to be produced. A simple presence/absence table of the type used in Harborne *et al.* (1971) is illustrated in Table 1. The more complicated Table 2 is the result of locating species containing at least one of three non-protein amino acids in the seeds and tabulating presence/absence of these amino acids with the species arranged by subgenera and

Table 1. Presence/absence table produced by the chemotaxonomic database.

Species	*Qr*	*Km*	*Fr*	*Cf*	*Le*	*Source*
L. amphicarpos	+	+	+	+	—	Ba81
L. angulatus	+	+	+	+	—	Pe59&To76
L. annuus	+	+	+	+	—	Pe59&To76
L. aphaca	+	—	+	+	—	Pe59&Ba81
L. cicera	+	+	—	—	—	Pe59&To76
L. cirrhosus	+	—	?	?	—	To76
L. clymenum	+	+	+	—	—	Pe59&To76
L. grandiflorus	+	+	+	—	—	Pe59&To76
L. heterophyllus	+	—	+	+	—	Pe59&Wi59&To76
L. hirsutus	+	+	+	+	—	Pe59&To76
L. incurvus	—	—	+	—	+	Ba81
L. japonicus	+	—	+	+	—	Pe59&To76
L. latifolius	+	—	+	+	—	Pe59&To76
L. laxiflorus	+	+	+	+	+	Ba81
L. linifolius	+	—	+	—	+	Pe59&To76
L. linifolius	+	+	+	—	+	Ba81
L. magellanicus	+	—	?	?	—	Wi59&To76
L. niger	—	+	—	—	—	Pe59&To76
L. nissolia	—	+	+	—	—	Pe59&To76
L. occidentalis	+	+	+	+	—	Pe59&To76
L. ochrus	+	+	—	—	—	Ba81
L. odoratus	+	+	?	?	—	Ha60&65&To76
L. palustris	+	+	—	+	—	Pe59&To76
L. pisiformis	—	+	+	—	—	Pe59&To76
L. pratensis	+	+	—	+	—	Pe59&To76
L. rotundifolius	+	+	+	+	—	Pe59&To76
L. sativus	+	+	—	—	—	Pe59&Wi59&To76
L. setifolius	+	+	—	—	—	Pe59&To76
L. sphaericus	—	+	+	—	—	To76
L. sylvestris	+	—	+	+	—	Pe59&To76
L. tingitanus	—	+	+	+	—	Pe59&To76
L. tuberosus	+	+	+	+	—	Pe59&To76
L. vernus	—	+	+	—	—	Pe59&To76

+ present, — not recorded, ? not surveyed.
Qr = Quercetin, Km = Kaempferol, Fr = Ferulic acid, Cf = Cafeic acid, Le = Leucoanthocyanidin.

Table 2. A list of *Vicia* species with one of three non-protein amino acids present in the seeds. The species are nested within subgenera (Vicia and Vicilla) and again within sections.

Subgenus/Section	*Canavanine*	*3-Cyanoalanine*	*4-Glutamyl- -3-Cyanoalanine*
Vicia			
Atossa			
sepium	not detected	present	present
Hypechusa			
hybrida	not detected	present	present
Peregrina			
michauxii	not detected	present	present
peregrina	not detected	present	present
Vicia			
grandiflora	not detected	present	present
lathyroides	not detected	present	present
pyrenaica	not detected	present	present
sativa	not detected	present	present
Vicilla			
Australes			
graminea	present	not detected	not detected
Cassubicae			
cassubica	present	not detected	not detected
Cracca			
benghalensis	present	not detected	not detected
caroliniana	present	not detected	not detected
cirrhosa	present	not detected	not detected
cracca	present	not detected	not detected
disperma	present	not detected	not detected
hirsuta	present	not detected	not detected
ludoviciana	not detected	present	present
monantha	not detected	present	present
palaestina	present	not detected	not detected
sicula	not detected	present	present
villosa	not detected	present	present
villosa	present	not detected	not detected
Ervilia			
ervilia	present	not detected	not detected
Ervoides			
articulata	present	not detected	not detected
Ervum			
laxiflora	present	not detected	not detected
tetrasperma	present	not detected	not detected
Panduratae			
cretica	present	not detected	not detected
Vicilla			
sylvatica	present	not detected	not detected

sections. It shows how the presence of one is correlated with the absence of the other two, and how this pattern cuts across the group of species known as section *Cracca*.

2. Data Applications

Selected subsets of the data can be stored in files within the computer and adjusted in format for entry to various taxonomic analysis programs. The linkage diagram in Fig. 1 shows the results of one single-linkage cluster analysis of *Lathyrus*, *Pisum*, *Lens* and *Vicia* spp. using data on phytoalexins extracted from the database in this

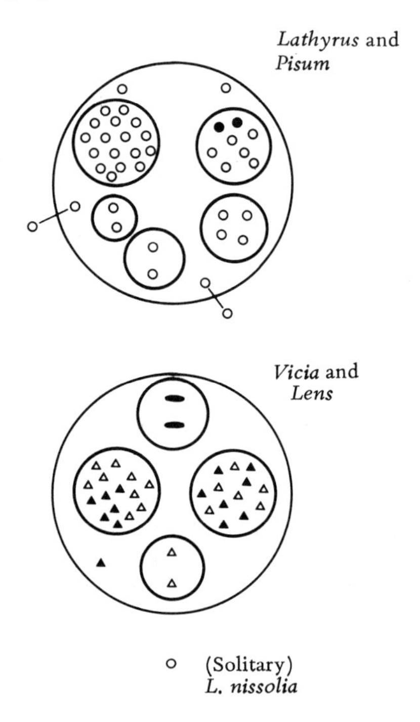

Fig. 1. A linkage diagram for Vicieae species linked by resemblance ⩾0.89 calculated from phytoalexin patterns extracted from the chemotaxonomic database. Each symbol represents one species. Species enclosed in a heavy circle are identical in phytoalexin response. Species or groups of species within a light circle have resemblances of 0.89 or greater with three or more other members of the circle. Species outside the circles are either linked by resemblance ⩾0.89 to only one species in a circle, shown by a line, or are isolated from them. ○, *Lathyrus* spp.; ●, *Pisum* spp.; ▬, *Lens* spp.; △, *Vicia* spp. (subgenus Vicilla); ▲, *Vicia* spp. (subgenus Vicia).

way. The two clusters shown (*Lathyrus* with *Pisum*, and *Vicia* with *Lens*) support the major division reported by Robeson and Harborne (1980), the authors of this data.

CONCLUSIONS

(1) Our chemotaxonomic database for the Vicieae contains data, where available, for 85 substances specified by organ and with reference citations for 111 species, just over one-third of the species.

(2) The database provides comparative chemotaxonomic data that can be retrieved in various forms useful to taxonomists, and be passed to other programs for taxonomic analysis.

(3) It proved feasible for one of us working full time to collate the data, perform some checking, comparison and infilling of data, design the database, and enter the data to a usable condition in 4 years.

(4) The database is now operated as part of the Vicieae Database Project and is used regularly to help with queries from the Study Group for that project.

ACKNOWLEDGEMENTS

We are very grateful to Professor E. A. Bell, Dr L. Fellows, Professor J. Harborne, and Dr J. Robinson for help, advice and the gift of standards.

REFERENCES

Abbott, L. A. (1976). "EXIR user's manual". University of Colorado, Boulder.

Bell, E. A. (1962). Associations of ninhydrin reacting compounds in the seeds of 49 species of *Lathyrus*. *Biochem. J.* 83, 225–229.

Bell, E. A. and Tirimanna, A. S. (1965). Association of amino acids and related compounds in the seeds of 47 species of *Vicia*: their taxonomic and nutritional significance. *Biochem. J.* 97, 104–111.

Bisby, F. A., Vaughan, J. G. and Wright, C. A. (eds) (1980). "Chemosystematics: Principles and practice". Academic Press, London and Orlando.

Bisby, F. A., White, R. J., Macfarlane, T. D. and Babac, M. T. (1983). The Vicieae Database Project: Experimental uses of a monographic taxonomic database for species of vetch and pea. *In* "Numerical Taxonomy" (J. Felsenstein, ed.), pp. 625–629. [NATO ASI Series G, Vol. 1.] Springer, Berlin.

Gibbs, R. Darnley (1974). "Chemotaxonomy of Flowering Plants", 4 vols. McGill-Queen's University Press, Montreal.

Harborne, J. B., Boulter, D. and Turner, B. L. (1971). "Chemotaxonomy of the Leguminosae". Academic Press, London and Orlando.
Pecket, R. C. (1960). The nature of the variation in flower colour in the genus *Lathyrus*. *New. Phytol.* 59, 138–144.
Robeson, D. J. and Harborne, J. B. (1980). A chemical dichotomy in phytoalexin induction within the tribe Vicieae of the Leguminosae. *Phytochemistry* 19: 2359–2365.
Smith, P. H. (1976). "The Chemotaxonomy of Plants". Arnold, London.
Torck, M., Bezanger-Beauquesne, L. and Pinkas, M. (1971). Recherches sur les flavonoids des Legumineuses: I. Etude Chimique. *Ann. Pharm. Fr.* 29, 201–210.

19 | BRASS BAND (The Brassicaceae Databank at Notre Dame): an example of Database Concepts in Systematics

T. J. CROVELLO

Department of Biology, The University of Notre Dame, Notre Dame, Indiana 46556, USA

L. A. HAUSER

Department of Botany, The University of Illinois, Urbana, Illinois 61801, USA

and

C. A. KELLER

Department of Education, Andrews University, Berrien Springs, Michigan, USA

Abstract: For over 15 years we have created, maintained, and even at times abandoned, computerized databases of the Brassicaceae. Review of these databases provides insight into the general development and evolution of databases in systematics, be they bibliographic, nomenclatural, herbarium, floristic, or descriptive. Data handling, the process of planning, creation, use and revision of databases is described as a multi-stage decision process. Specifically, valuable systematic databases require many steps and demand that decisions be made at each stage. The *wise* use of computers can enhance systematics, but in the past too many systematists have assumed that if a computer can do something, then it is worth doing, and is best done by computer. Over the years systematics has

Systematics Association Special Volume No. 26, "Databases in Systematics", edited by R. Allkin and F. A. Bisby, 1984, pp. 219–233. Academic Press, London and Orlando.

DATABASES IN SYSTEMATICS
ISBN 0 12 053040 6

suffered from a lack of coordination among institutions. Even widely publicized sensible guidelines are not followed. Many earlier problems should disappear as more systematists develop increased familiarity with computers and as database management systems become available to systematists. Such systems will allow smooth transition from one task to another, including information retrieval, data analysis, graphics and word processing.

INTRODUCTION

Computerized database projects in systematics have increased in number and diversity over the last 20 years. Early pioneers included Gould (1958, 1963) who developed codes for plant taxa and whose work met great resistance from systematists. Perrring and Walters (1962) and Soper (1964) used computers to map plant distributions. MacDonald (1966) developed computer methods for botanical garden and arboretum records. Crovello (1967) summarized a survey of museum curators and problems in the use of electronic data processing in biological collections. Perhaps Notre Dame's Edward Lee Greene Herbarium with its 65 000 specimens was the first non-trivial collection to be completely computerized (Crovello, 1972). It still serves the systematics community via customized computer searches. Systematic databases were also the basis of new analytical approaches collectively referred to as numerical taxonomy (Sokal and Sneath, 1963).

Today the computer permeates all facets of society, science and systematics. Databases in systematics are being developed in large numbers by individual systematists and institutions (e.g. see Sarasan and Neuner, 1982). Twenty years ago controversies and problems ranged from whether computers would replace curators, through ignorance about computers, to the non-existence of necessary equipment, its cost and its availability to systematists. Today computers are an accepted tool in systematics. Current problems involve not the non-existence of adequate computers but the existence of too many types, and of too many procedures to create systematic databases. Unfortunately, many current projects are concentrating on how best to computerize, without first asking such important questions as: is this particular project really worth computerizing; and are the resources for it to reach a valuable stage certain? Because such questions were ignored, many once promising projects are now extinct.

Our first goal is to describe BRASS BAND, a computer-based BRASSicaceae dataBank At Notre Dame (as is customary, acronyms of databases or programs will be in capitals throughout this paper). Since BRASS BAND includes bibliographic, nomenclatural, herbarium, floristic and descriptive databases, it also serves our second purpose: to characterize trends in systematics databases over the past 15 years. Finally, we will emphasize that creation, maintenance and use of databases should be viewed as a multi-stage decision process.

COMPONENTS OF THE BRASS BAND

The following paragraphs describe each database that was, is, or will be found in the BRASS BAND. Presentation of these diverse databases indicates the diversity of decisions and traits involved in data handling. A review of these databases should also emphasize that any serious databanking project is long term, evolves in often unexpected ways, and uncovers both problems and uses that were not anticipated.

1. The LIT BRASS Database

Around 1970 Crovello created a bibliographic database to accommodate articles and books on a variety of subjects relevant to his interest. These focused on *taxa* (e.g. Brassicaceae, Culicidae and Salicaceae), *topics* (e.g. geographic variation, speciation and economic plants), and *techniques* (e.g. numerical taxonomy and chromatography). A set of about 200 descriptors were defined. Besides taxa, topics and techniques, additional descriptors included geography, morphology, physiology and whether a reprint was available at Notre Dame. By 1974, over 7000 references were in the literature bank. The program was a modification of that used for specimen label data in the Greene Herbarium project described below. By 1976, Crovello decided to abandon maintenance of LIT BRASS and the entire literature database because of the time taken to index each reference and record the information; and the appearance of a more cost effective, up to date substitute, namely commercially available on-line information retrieval, especially using BIOSIS. Since searches on taxon names are simple, an annual search for all references relating to the family or several genera takes only minutes and costs about $25 to $50. Because the taxonomic literature on the family is not that extensive (excluding distribution records), the resulting print-outs serve his needs.

2. *The INDIANA BRASS Database*

The State of Indiana, approximately 93 000 km^2 in size, is located in the midwestern United States. Over the years, several separate databases have developed in INDIANA BRASS.

Crovello and Keller (1974) created a floristic database summarizing our knowledge of the distribution of 81 taxa of the Brassicaceae in each of Indiana's 92 counties. Both the scientific and common name of each taxon were entered into the University's maxicomputer. Five PL1 programs were written to store the data on magnetic tape, to produce a map of Indiana with county outlines and values such as number of species, to shade each county differentially depending on its value, to calculate a divergence value between counties, and to produce county checklists. The original database was also input to NTSYS, a set of numerical taxonomy programs developed by F. J. Rohlf and colleagues, now at SUNY, Stony Brook, New York. This may be the earliest attempt to calculate *phyto*geographic divergence among geographic units, to create a phenogram of their relationships, and to summarize them on a map using a technique like differential shading. At about the same time, Sneath and McKenzie (1973) ingeniously presented the results of an ordination of conifer taxa on a geographic map of the world.

Keller (1979) later studied the distribution of all of the approximately 2200 vascular plant taxa of Indiana. Using innovative methods he summarized similarities among Indiana's 92 counties and grouped them into six floristic areas. He created a quantitative index to optimize assignment of each of the 92 counties to one of the six areas. He wrote a series of PL1 programs to perform error checking and analyses of this large database.

Since Keller's database included the scientific name of each taxon, it formed the basis of an updated computer-based check-list of Indiana's vascular plants (Crovello, Keller and Kartesz, 1983). Again, error checking and production programs were written in PL1. It proved extremely tedious to proof-read such an extensive list, and other problems arose. The first was its long production time: over six years of regularly interrupted work. This wasted considerable time remembering where we had stopped, which version of the check-list had which corrections, etc. The available computer hardware also changed, e.g. both upper and lower case characters became available. Keller wrote an additional program to convert all names to upper and lower case yet not violate the rules of nomenclature. Finally, a check-list for North America north of Mexico appeared (Kartesz and Kartesz, 1980). We now felt the need to relate our check-list not only to the standard Flora of the State but also to this more recent check-list.

3. The NORTH AMERICAN BRASS Database

In collaboration with John Kartesz of the Biota North America Project, Crovello is creating a computer database of the distribution of the approximately 650 taxa of the Brassicaceae found in North America north of Mexico. Data are based on the taxonomic literature. The resulting matrix serves as input to the NTSYS program package, to the SAS and SAS/GRAPH packages (Anon., 1979, 1980) and to SYMVU, the three-dimensional geographic-based plotting program available from Harvard University's Laboratory for Computer Graphics.

4. The EUROPEAN BRASS Database

Crovello created a database for Europe similar to NORTH AMERICAN BRASS. *Flora Europaea* was used to obtain the occurrence of each taxon among the countries of Europe. He plans analyses similar to the above.

5. The SOVIET BRASS Database

Crovello and Miller (1982) summarized several analyses of a database of the 736 species of Brassicaceae distributed among the 51 geographic regions of the Soviet Union. For each species they also captured its binomial, species number, year the species was published, general world distribution, habitat and months of flowering. All the above data were initially entered for use by FLEUR, a powerful, general-purpose information retrieval program, written in PL1 by Douglas Miller while an undergraduate at Notre Dame. Other uses of FLEUR have involved literature references, photographic slide inventories, etc. A current use involves the preparation of bibliographies of thousands of literature references for The North American Benthological Society. FLEUR output was modified by R. A. Hellenthal of Notre Dame to produce input in a format acceptable to SCRIPT, a word-processing language available on IBM maxicomputers. SCRIPT's output is camera-ready copy.

For SOVIET BRASS the FLEUR program quickly retrieved lists such as those few species restricted to the Arctic area, or species in the Soviet Union described by Komarov. More difficult Boolean logic searches are also possible. For example, which species are in Central Asia and the Arctic, but not in the Caucasus?

The 736 species by 51 OGU (Operational Geographical Unit; see Crovello, 1981) database also served as input for analyses using SAS, NTSYS and WAGNER78, a cladistics program written by J. S. Farris (SUNY at Stony Brook, New York). Three-dimensional *biotopomaps* (Crovello, 1981) were produced using SYMVU, while BIOPT4 (K. Grant, Notre Dame) produced three-dimensional graphics of results from Principal Components Analysis.

Creation and use of SOVIET BRASS by such a diversity of programs represented several important new approaches to the use and maintenance of systematic databases. It provided several capabilities: to retrieve selectively subsets of information from a database; to use such output as direct input to programs for multivariate analysis or to produce sophisticated graphics; and to pass prose to word-processing programs that can produce camera-ready copy or drive typesetting machines; and all of this with minimal repeated data capture.

During creation of SOVIET BRASS our views on maintenance and editing programs changed. Originally we planned to create a separate program to update our FLEUR databases, but realized that many desired features were already available in the EDIT mode of Notre Dame's IBM maxicomputer. This program had several advantages: it is well known; manuals and consultants are available; it has many more capabilities than we might write; is periodically updated; and has required no personnel or expense to create or maintain it.

6. *The* Arabidopsis *Database*

Arabidopsis is a genus of small plants native to Europe. *Arabidopsis thaliana* (L.) Heynh. is a cosmopolitan weed used in physiological and genetic studies because it can complete its life cycle in 14 days, in a test tube, on defined nutrient media. Crovello (1968) created a database of 154 genetic lines from 120 strains of the species. Data on 12 characters were obtained from Röbbelen (1965) and, with his permission, were analysed using standard phenetic techniques available on NTSYS.

7. The Greene Herbarium Database

Edward Lee Greene (1843–1915) is an important and controversial figure in North American systematics. He named or recombined over 4400 taxa and amassed an herbarium of 65 000 sheets. Unfortunately, he left no notebooks to help later generations understand his decisions or locate his type localities. In 1968, Crovello received a National Science Foundation grant to develop a database of label, annotation and accession information of the 65 000 sheets in the Greene Herbarium. Limited "state of the specimen" information was also recorded. Details of the information and programs used to create, check and query the database can be found in Crovello (1972). The project took three years to complete and involved Crovello, about six months of a programmer's time (an inexperienced computer science graduate), three years of an intelligent secretary who could accurately code and record information directly from herbarium labels, and an undergraduate to make production runs on the computer. The project is characterized by the large amount of information captured per sheet, the taxonomic value of many sheets (e.g. type material), and by offering customized computer searches since about 1970.

8. The Cardamine *Database*

To begin a monographic study of New World *Cardamine* and related genera, Crovello borrowed about 31 000 specimens from about 70 herbaria. After verifying the species to which each belonged, similar data were recorded for each specimen as in the Greene Herbarium project. Additions included Index Herbariorum codes.

Since only about 200 taxon names were involved, a code consisting of a letter (for each genus) and a number (for each species in a particular genus) was used when recording original identifications or subsequent annotations. For example, C101 represents *Cardamine bellidifolia*. The advantages of coding are significant reductions in keystroking and proof-reading. Also, since the code is retained in the database, users have the option of using them instead of typing the actual names. A serious drawback of such coding is that data errors are difficult to detect. While an error-checking program was used to discover invalid codes, only time-consuming comparison of a print-out with the identification of each specimen could uncover the use of codes that are valid, but incorrect, for a given specimen.

In such cases a cost/benefit analysis should be considered to choose among alternative procedures. These include using no abbreviations or codes; using only those with some built-in redundancy; or using those without redundancy. Alternatively, coding might be restricted to those properties that are less important than the scientific name. The cost/benefit calculation itself would include information on the rate of errors and the total number of specimens entered into the database. If the latter is great, then inter-specimen redundancy might offset the risks of lack of redundancy in a code. For example, if a specimen were incorrectly coded, it might be revealed when the geographic positions of all specimens of the species to which it was incorrectly assigned are plotted on a map. If it is not an outlier, the effect of such a coding error on any taxonomic conclusion may be small enough to ignore. Our point is that a totally error-free database may be impossible to attain, or the level of accuracy required by the database's uses may not justify the extra expense to make it absolutely error free. The example given also indicates that redundancy at one level of the database hierarchy can compensate for errors at another.

9. The Thelypodieae Databases

The Thelypodieae may be the most primitive tribe in the Brassicaceae. Nine of its genera are North American, while the tenth (*Macropodium*) occurs in southern Siberia and the Soviet Far East. A series of databases were constructed for phenetic, cladistic and biogeographic analyses.

Hauser and Crovello (1982) based their analysis of inter-generic relationships on a 10-genus by 32 morpholoigcal character database. The adequacy of a single character state representing a taxon decreases as the taxa being analysed are located higher up the taxonomic hierarchy. The character state used in each cell of the 10-genus by 32 character database is the predominant state within the genus. If one were not clearly dominant, the genus was coded as variable. Measurements were taken from herbarium specimens, with a few from the literature.

In a much more extensive study of the tribe Hauser (1982) created the following databases: 93 species by 33 morphological characters; 93 species by 75 OGUs covering the western United States and Mexico; and six species by morphological databases, one for each of the six nonmonotypic genera of the tribe. Among these last six the number of species ranged from 4 to 34, and the number of characters from 17 to 75.

Interestingly, each of the above databases used different character sets to some degree. No one character set was found that could simultaneously discriminate within each genus and among genera.

10. Intergeneric Databases across the Entire Family

Keller and Crovello (1973) developed procedures to transfer data efficiently from floras and monographs into a computerized database. As an example, they used descriptions of genera of the Brassicaceae found in three Floras: two from eastern North America (Fernald, 1950; Gleason, 1963); and one from Europe (Tutin *et al*., 1964). Briefly the method was: develop a list of characters and character states with an abbreviated code for each, transform each datum in a Flora (in keys as well as in prose descriptions) to the code that describes the state of each character in each genus; enter the coded data into the computer; transform coded data from keys into the same format as used in prose descriptions; rearrange all transformed data into simple "card image" format; integrate data from different floras into one data set sorted by character and character state codes; produce integrated descriptions of any desired set of taxa, e.g. species in Indiana; create a character by genus data matrix for quantitative analysis; and edit this matrix.

As is evident from the many complex steps just listed, to transform prose and key data into a descriptive database is not a simple task. Other problems encountered were not related to computing. For example, although Fernald (1950) used 127 different characters to describe at least one of his 43 genera, and Tutin *et al*. (1964) used 144 to describe at least one genus, the average number of characters used to describe each genus in their works was under 30. Consequently, a genus by character database with extensive missing data results. Another problem is disagreement between different literature sources. Both may be correct if they describe different species within the genus and a range of values is appropriate. But data formats for multivariate analyses do not accommodate within taxon variation. Conversely, if two Floras treat the same species and disagree, further study is necessary.

Consider also the problem of modifer words in descriptions (e.g., *somewhat* pubsecent; *rarely* over 10 cm), of vague statements (e.g. "south to Georgia and Mississippi"); of "not" phrases in keys (e.g. "without combination of characters described above"); and of different authors' concepts of what each taxon encompasses. All of these, but especially the last, can be particularly serious at the species level and below.

11. Useful non-biological Databases

Systematic research will always require geographic and ecological information. Databases in these areas are especially valuable because systematists are less familar with the literature in these fields. The GEOECOLOGY database (Olson *et al.*, 1980) was an important milestone, providing information at the United States county level on climate, vegetation, physiography, etc. available on magnetic tape compatible with the SAS program package. Thus systematists not only have a large, valuable, proof-read data set, but do not need to write extensive information retrieval programs to access it!

DATABASE TRENDS IN SYSTEMATICS

Consideration of the characteristics of databases which systematists began, completed or abandoned over the last 20 years can assure better decisions in the future. Due to space limitations we can do little more than mention some important traits and trends in the following paragraphs.

Too often a systematics database project begins with no formal, written, *a priori* statement of goals, procedures and evaluation of resources. Yet such a step is essential to determine if the project is worth it to systematics, and to the institution and systematist involved. Too many projects have failed from lack of commitment by higher administrators. Not only would a formal *a priori* presentation allow administrators to assess the value of such a project but, if approval is forthcoming, it will likely be a firmer, long-term commitment.

Project leaders frequently fail to identify the target audience, and its size. They simply assume that creation is academically, if not commercially, valuable. What must be determined is whether the same resources could better serve other uses. Related to this is the frequent underestimate of time, finances, space, etc. required. Larger projects require more time per unit record, because of increased file management, etc. Also, in any project a minimum amount of resources *per unit time* must be devoted to the project to maintain project momentum and minimize time lost trying to remember what files have been checked, corrected, etc.

Too many systematists have assumed that if a computer can do something then it must be worth doing, and that it is best done using a computer. These assumptions are dangerous. Others assumed that it is valuable to put all available data in a database. Label data from all herbarium specimens should not be computerized unless it can be justified (cf. the Greene Herbarium project). Among those specimens or records thought to be worth capturing, the most valuable data should be computerized first.

Over the years the field of systematics has suffered from lack of coordination among database projects. Important reasons include lack of funding coordination, need to use the computer available at each location, different purposes for the different projects, and differences among project leaders about what information is to be stored. Sensible guidelines developed by groups like the Computer Standards Committee of The Association of Systematics Collections have been mostly ignored. The growing availability of diverse micro-computer systems may aggravate this problem.

Multi-purpose databases are becoming the goal and reality in systematics, i.e. many systematists are now thinking of a multi-faceted database management system that can accept many diverse databases, search for relevant data, merge the results, carry out routine statistical, or special analyses, produce summary graphics, and even interface with a word processor or computer typesetting machine to produce publication copy. While perhaps no systematist has the ideal database management system at his disposal, many are able to avoid major repeated data capture between stages. The increased availability of powerful, multi-task program packages like SAS, and their ability to create output files to serve as direct input to systematics applications programs like NTSYS and WAGNER78, will help us achieve more effective database management.

A final consideration is whether it really is worth the effort to

Table 1. Data handling in systematics: a multi-stage decision process (return to preceding stages is frequent).

1. Purpose of the database project.
Create a detailed, *written* statement of general and specific purposes of the project. Include its value to science, society, the institution, the principal invesigator, etc. Indicate the services to be supplied and who will be served. Estimate if the intended audience really wants the services (enough to buy them?).

2. Basic decisions and activities.
Choice of taxa, populations, specimens, characters, character states, geographic limits, etc. Choice of computer system (hardware, software, people). Adequacy of available computing systems. Estimation of required resources, including all hidden costs. Sources of funding. Tentative timetable. Obtain counsel of those with database experience and of potential users. Develop reasons why project can benefit from computers.

3. Proposal presentation.
To one's institution or outside funding sources. Incorporate suggestions for improvement.

4. Proposal endorsement.
Obtain *written* promise of funding and assurance that one's institution sees this project as an important part of the systematists' job activity.

5. Use of previous information.
Literature retrieval. Actual data (field data; living collections; preserved collections; experimental data; abiological data).

6. New data measurements.
(as in item 5).

7. Data entry.
Assemble actual data sources. Transcribe or encode data. Enter and verify data.

8. Database creation.
Automated error checking. Multiple file creation.

9. Data retrieval.
Choice of search parameters, output format, etc.

10. Data analysis.
Prose or tabular results; simple statistics; cluster analysis; ordination; geographic maps; numerical cladistics or biogeography; etc.

11. Systems analysis.
Databases as sources of raw data for simulations.

12. Decision making.
Databases and analyses to aid in classification or identification.

13. Publicizing results.
Conventional publication. Deposition of new results back into the database. Word processing.

14. Maintenance and growth of the dynamic databank.
Corrections; additions; deletions; adding databases created at other institutions.

merge systematic databases. In some cases the answer is obviously "yes", e.g. herbarium databases of several institutions, but even then an *a priori* formal analysis should be made. Factors to consider include amount of information captured per specimen, accuracy of the information, the taxa and geographic areas covered, ownership and royalties, etc.

DATA HANDLING IN SYSTEMATICS: A MULTISTAGE DECISION PROCESS

From the database examples presented in this and other papers, it is clear that *data handling*, the process of planning, creation, use and revision of databases, is a multi-stage process. Table 1 summarizes its major stages. Unfortunately, space limitations prevent us from discussing it in detail, but we emphasize that its many stages and substages require many decisions, which affect the success or failure of the project, concerning how well it serves the needs of systematists and others. In this way data handling in systematics is similar to any study in ecology or systematics (Crovello, 1970) or in biogeography (Crovello, 1981).

CONCLUDING REMARKS

Our review of the diverse databases in BRASS BAND to some degree reflects the general evolution of databases in systematics. Today, a sufficient number of systematists have considerable expertise with computers, and not just for information retrieval. More importantly, they realize that computers are not a panacea and that they require considerable resources. Systematists today are less motivated to create computer databases for superficial reasons. We can expect that databases that are created will be used much more intensively and in more ways than in the past. The effect of microcomputers is still difficult to predict, but they certainly must be considered. Our guideline should be that the *wise* use of computers can enhance systematics.

REFERENCES

Anon. (1979). "SAS User's Guide". SAS Institute, Raleigh.
Anon. (1980). "SAS/GRAPH User's Guide". SAS Institute, Raleigh.

Crovello, T. J. (1967). Problems in the use of electronic data processing in biological collections. *Taxon* **16**, 481–494.
Crovello, T. J. (1968). The value of numerical taxonomy for *Arabidopsis* researchers. *Arabidopsis Info. Serv.* 5, 7–9.
Crovello, T. J. (1970). Analysis of character variation in ecology and systematics. *Ann. Rev. Ecol. Syst.* 1, 55–98.
Crovello, T. J. (1972). Computerization of the Edward Lee Greene Herbarium at Notre Dame. *Brittonia* **24**, 131–141.
Crovello, T. J. (1981). Quantitative bigeography: an overview. *Taxon* **30**, 563–575.
Crovello, T. J. and Keller, C. A. (1974). Uses of computerized floristic data of Indiana for plant geography. *Proc. Indiana Acad. Sci.* **83**, 399–406.
Crovello, T. J. and Miller, D. C. (1982). Floristic similarities among 51 regions of the Soviet Union based on the Brassicaceae. *Taxon* **31**, 451–461.
Crovello, T. J., Keller, C. A., and J. T. Kartesz (1983). "A Computer Based Checklist of the Vascular Plants of Indiana". University of Notre Dame Press, Notre Dame.
Fernald, M. L. (1950). "Gray's Manual of Botany" (8th edn.). American Book Co., New York.
Gleason, H. A. (1963). "The New Britton and Brown Illustrated Flora". Hafner, New York.
Gould, S. W. (1958). Punched cards, binomial names, and authors. *Am. J. Bot.* **45**, 331–339.
Gould S. W. (1963). International Plant Index: its methods, purposes, and future possibilities. *Taxon* **12**, 177–182.
Hauser, L. A. (1982). Quantitative Phylogenetic and Phytogeographic Studies in the Thelypodieae (Brassicaceae). Ph.D. thesis, The University of Notre Dame, Indiana.
Hauser, L. A. and Crovello, T. J. (1982). Numerical analysis of generic relationships in the Thelypodieae: Brassicaceae. *Syst. Bot.* **7**, 249–268.
Kartesz, J. T. and Kartesz, R. (1980). "A Synonymized Checklist of the Vascular Flora of the United States, Canada, and Greenland". University of North Carolina Press, Chapel Hill.
Keller, C. A. (1979). Quantitative Techniques for the Determination of Phytogeographic Areas. Ph.D. thesis, The University of Notre Dame, Indiana.
Keller, C. A. and Crovello, T. J. (1973). Problems and procedures in the use of floristic data in a computerized data bank. *Proc. Indiana Acad. Sci.* **82**, 116–122.
MacDonald, R. D. (1966). Electronic data processing methods for botanical garden and arboretum records. *Taxon* **15**, 291–295.
Olson, R. J., Emerson, C. J. and Nungesser, M. K. (1980). "GEOECOLOGY: A County-level Environmental Data Base for the Conterminous United States". [Environmental Science Division Publication No. 1537.] Oak Ridge National Laboratory, Oak Ridge.
Perring, F. H. and Walters, S. M. (1962). "Atlas of the British Flora". Thomas Nelson & Sons, London.
Röbbelen, G. (1965). [Strain variation in *Arabidopsis thaliana* (L.) Heynh.]. *Arabidopsis Info. Serv.* 2, 36–47.

Sarasan, L. and Neuner, A. M. (1982). "Museum Collections and Computers". Association of Systematics Collections, Lawrence, Kansas.
Sneath, P. H. A. and McKenzie, K. G. (1973). Statistical methods for the study of biogeography. *In* "Organisms and Continents through Time". (N. F. Hughes, ed.), pp. 45–60. The Paleontological Association, London.
Sokal, R. R. and Sneath, P. H. A. (1963). "Principles of Numerical Taxonomy". Freeman, San Francisco.
Soper, J. R. (1964). Mapping the distribution of plants by machine. *Can. J. Bot.* 42, 1087–1100.
Tutin, T. D., Heywood, V. H., Burges, N. A., Valentine, D. H., Walters, S. M. and Webb, D. A. (eds). (1964). "Flora Europaea", vol. 1. Cambridge University Press, Cambridge.

20 An Outline for a Database within a Major Herbarium

J. M. MASCHERPA and G. BOCQUET

Conservatoire et Jardin botaniques, CH-1292-Chambesy, Geneva, Switzerland

Abstract: Besides its obligation to maintain and enlarge its collections, an important municipal herbarium has to play a role in the life of the city. The position of the museum in the local community implies a responsibility towards environmental problems including the urban and suburban zones, the history and inventory of gardens, parks and properties. A new floristic undertaking is thus proposed, in which the computer and databases take the place of conventional catalogues and books, allowing a "permanent revision" of the flora.

A data-processing system for our museum has been initiated as a result of these considerations. There are three outstanding needs: management, research and provision of information. As far as the management of the library and the herbarium is concerned, we distinguish administrative work (catalogue, loans, exchanges) from the purely scientific work (inventories, cross-checked lists, cartography). The research is related to the scientific management of the collections. It is based on the data accumulated in the collections, from which new information can be extracted. A better knowledge of the collection, and thus a better understanding of nature, should result. This depends on regular data revision and the diffusion of information through popular and university teaching. Each of these subjects has led to a particular study, resulting in the use of computerized techniques, some established and some new (including teleprocessing, databanks, text or image processing), in a coherent integrated system. Examples are given of the Institute's projects – a regional Flora, Latin-American and Mediterranean floristic studies.

Systematics Association Special Volume No. 26, "Databases in Systematics", edited by R. Allkin and F. A. Bisby, 1984, pp. 235–248. Academic Press, London and Orlando.

DATABASES IN SYSTEMATICS
ISBN 0 12 053040 6

INTRODUCTION

Apart from a few well-known cases (Crovello, 1972; Hall, 1974; Forero and Pereira, 1976; Morris and Manders, 1981), the major herbaria have often backed down when faced with the enormity of the problems to be solved and with the expense of placing records of large collections on computer. On the other hand, many projects involving fewer than 100 000 specimens have been started during the last 20 years (cf. the compilations of Crovello and MacDonald (1970) and Brenan *et al.* (1975) or the articles of Gómez-Pompa and Nevling (1973) and of Sweet and Poppleton (1977)). We have already discussed the advantages of computerized collections (Bocquet and Mascherpa, 1982): easy retrieval of raw data for preparing inverted lists, for personalized inventories by collector, locality, habitat or species, for distribution maps, and for automatic identification. For these reasons, it was decided in 1979 to introduce a computer to our Institute and to direct research towards computerization.

The Conservatoire and Jardin botaniques are financially dependent on the town of Geneva under the supervision of the Department of Culture and Fine Arts. Finance is thus guaranteed by the Geneva municipal community which, for a town of 150 000 inhabitants, devotes a considerable effort to the maintenance of local culture and tradition. Alongside its obligation to preserve and increase the collections, our Institute is therefore obliged to play an important part in the life of the local community and must assume responsibility for environmental problems of the urban zone, the suburbs, and the inventory of gardens, parks and properties. Furthermore the international reputation of our collections developed by De Candolle, Boissier, Burnat and their successors, leads many foreign scientists to visit our Institute each year. This dual local and international function led us to develop a computer system designed for our museum's particular position. We need to be able to answer the questions both of experienced research workers and of the Geneva public. This has, therefore, led us to consider three main requirements: management, research and information.

To have the slightest chance of success in changing our system, it was necessary to design a project which would in no way upset the present functioning of the Institute nor require special investment at the start. It was also necessary to study, in as realistic a manner as possible, the various computer-oriented solutions. To avoid undertaking a project so large that it would never be finished and too small to uncover all the problems, we launched a project of medium size, managed by four to five researchers.

It was essential, furthermore, that this pilot project include all the Institute's activities: herbarium, library, research, garden and even popularized information. The project had also to be sufficiently original to obtain research funding and so on. We aimed at the Institute's traditional specialist interests: the local flora, the Mediterranean and Paraguay (Bocquet and Mascherpa, 1982).

COMPUTER SYSTEM

1. Hardware

Because of an agreement, in 1979, between the City and the State of Geneva, our Institute enjoys university status and we may use the facilities of the University's Computer Center (UCC). Consequently, we are members of a network centred on the mainframe computer UNIVAC 1100/61 (756K memory). We have two on-line (9600 bds) full-duplex displays, as well as a printer-plotter PRINTRONIX.

To replace the collection of ageing machinery which we used to publish the *Candollea*, *Boissiera*, *Saussurea* series and numerous other works, we acquired a text-processing and phototypesetting system AUTOLOGIC, based on an APS1000 control unit (256K bytes and 2 floppy discs 2D/2S). This system allows several SCRIB units for text capture to be connected. It is linked to the UCC's computer by a TTY line, which renders possible the production of phototypeset texts from various databases by suppressing the publication of the actual computer listings (Fig. 1). Example output from this system are found in Bertoni *et al.* (1982) and Encarnación *et al.* (1982). A special feature of our system is the ability to move directly from computer listings to camera-ready copy.

To offer an extra service to our library visitors, we have for the past three years set aside a visual display and printer for link up with American and European bibliographical databases. The richness of our specialist library is such that references obtained by computer are usually found on its shelves.

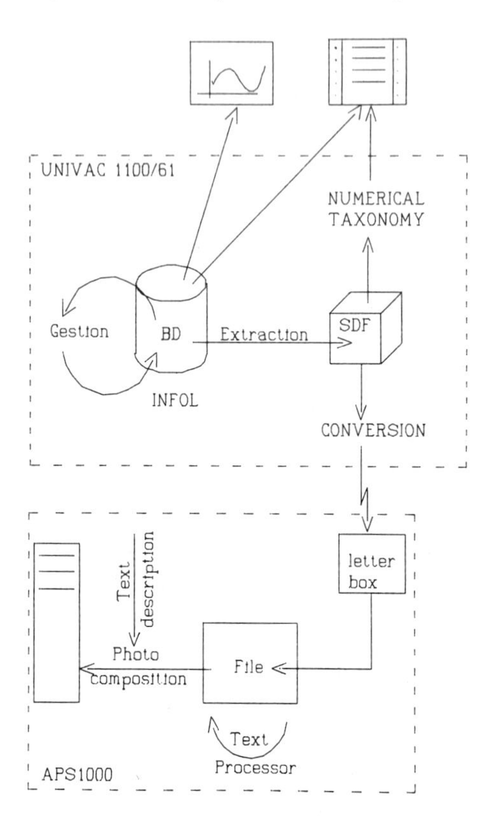

Fig. 1. Links between files stored on UNIVAC 1100/51 and APS1000.

2. *Software*

(a) Management. We distinguish the purely administrative tasks from the processing of information. Management of the collections is linked, as far as possible, to systems already in existence or whose development only requires limited computational effort. Thus, for example, we did not include herbarium loan management. In the future, herbarium exchange and new acquisition records will be computerized, to achieve a management system for records of all material received or on loan.

The library is soon to be linked to the REBUS network (REseau des Bibliothèques Utilisant SIBIL) started some ten years ago by the

Library of the University and Canton de Vaud, in Lausanne. The SIBIL package (Système Intégré pour les BIbliothèques de Lausanne) is superbly adapted for running a large library (Gavin *et al.*, 1980). Their main aims being efficiency and rationalization, the Swiss French-speaking university libraries decided to set up a common network based on the IBM 4341 computers of the Lausanne library and the computer centre of the State of Vaud. A common computerized catalogue will save a lot of precious cataloguing time. Automatic editing and the distribution of microfiche from this catalogue with on-line inquiry available in major libraries will facilitate both access to documents and inter-library loans. It is also planned to extend the system used in Lausanne for the management of loans, reminders and orders to the entire network.

(b) Research. Of the Institute's research programmes (Bocquet, 1982), three directly involve computerization.

A *floristic and phytosociological database for Corsica* is being set up. Specialist projects also centre on the distribution of various Mediterranean species, e.g. the genera *Silene* and *Digitalis*. In collaboration with the Botanical Garden of Berlin-Dalhem (FRG) and the Ecothèque Méditerranéenne de Montpellier (F), a *Mediterranean basin floristic list* (Med-checklist) is being completed. Extraction and analysis of the information contained in the literature is carried out in Berlin-Dalhem under the supervision of W. Greuter, who also has responsibility for all taxonomic and nomenclatural decisions. The Geneva group (H. M. Burdet and co-workers) search the literature, check nomenclatural details (authorities, dates, etc.) and keep bibliographic records. Computerization and the database are entrusted to G. Long at Montpellier (Greuter *et al.*, 1981). The computer lists produced in Montpellier, and the bibliographical catalogues from Geneva will be automatically edited using our text-processor.

South-American studies, centred for the most part on Peru and Paraguay, also involved a computer system (Spichiger, 1982). They needed to collect new data and undertake theoretical and practical studies for the creation of a Flora of Paraguay (Bertoni *et al.*, 1982). A flora of the woody species of the Peruvian Amazon has been initiated with the support of Swiss Technical Cooperation and a floristic inventory and a description of the vegetation in various parts of the Peruvian Amazon have been completed with the aim of producing a computer-stored Flora. One of the first tasks was to create a bibliographical database for this region which has been published as a list with keyword comments (Encarnación *et al.*, 1982).

In *local floristic studies*, a plan for a continually revised Flora of the Geneva area is central to our feasibility tests for computerizing our Institute. Bearing in mind the special position of our Institute, we have tried to tackle the main problems associated with the environment, urban plantations and the suburbs. Furthermore, we have given prime importance to the need of our students to integrate with town life, by association with Swiss professional organizations and by suggesting practical research projects.

In the Parks and Gardens project, we created a research link between our Institute (specialized in the theoretical problems of taxonomy and nomenclature) and the municipal service "Parcs et Promenades" which resulted in a thesis by a forestry engineer. The project involved the management of a living collection of trees using up-to-date solutions to the problems of inventory and mapping, as well as responding to the questions from the general public. Although of less scientific importance, it is worthwhile remembering the political importance of questions which affect our environment and that the Geneva Municipal authority has just granted two million Swiss francs for, amongst other things, restocking the tree population of our city.

A study of the weeds in the Geneva area, in association with the Federal Center for Agronomic Research in Changins, has resulted in a thesis by an agronomic engineer. In highly agricultural cantons such as Geneva, the combination of technical and practical expertise in research aimed at day-to-day agricultural problems with theoretical systematic research in the field, herbarium and library, should lead to studies of the development of our Flora. A conventional floristic approach is also necessary combining inventories, verified in detail, obtained from the literature and herbarium, with systematic searches of new sites, particularly road-sides: indispensable information for our continually revised local Flora.

A final task remained: the creation of a computerized catalogue of the Botanical Garden. We have been helped in our task by the fact that the Geneva Botanical Garden did not have a centralized catalogue, but a selection of small catalogues spread among various departments (Hunt, 1977). A central catalogue was conceived in 1981 and begun in 1982 (Iff, 1982) The management of new arrivals (of living material, seeds or dry specimens) is completely computerized. The system also automates the addressing of envelopes and production of the seed catalogue, and in the near future will manage botanical garden exchanges and the automatic production of garden labels. Annual review of the statistics will make it possible to document exchanges. Species identification and automatic mapping of cultivated areas have also been studied. A special grant has just been agreed by the Geneva Authorities to establish a grid for all the town's parks based on the national coordinate network. The management of Geneva's parks is a joint effort.

PACKAGES

1. INFOL Language

At this stage in our work, we have not wished to restrict future development by deciding upon one data structure or by choosing one of the commercially available high-level packages. For example, while it is true that systematic information is organized hierarchically, it is by no means certain that a hierarchical tree database model is best adapted for research in herbaria or libraries. Indeed, the relationships existing between taxa are ineffectively expressed. On the contrary, to build a database on the flat model is not the best way of expressing the hierarchical structure of taxonomic information. Network structures have advantages over both structures described above, but their disadvantages become apparent when implementing the various logical routes to be followed. It appeared to be more judicious to study in depth the structure of botanical information itself and of all possible networks. It was necessary to design simple recording sheets, both for specimens already in the herbarium and for new collections in the field, and to use a traditional program package for information storage and retrieval. We also wished to test data capture in the herbarium, library or field, using staff not familiar with computers, before finally deciding upon a work plan.

In this pilot project, we chose to use the INFOL language (INFormation Oriented Language) conceived around 1965; it was designed in 1968 by Northwestern University (Evanston, Illinois), and further developed at Geneva University and CERN (European Center for Nuclear Research). Two versions of this language exist, one for batch processing and the other for interactive use (Levrat *et al.*, 1979).

INFOL is a sequential language based on the "element" notation; the information concerning each element or object is separated into characters ("items"). The language comprises four phases:

(1) an ESTABLISHMENT phase in which the structural information is defined, the characters described, and the descriptor types defined (numerical, alphanumerical, code or date); this phase includes initial data capture (Fig. 2);
(2) an UPDATE phase during which data may be corrected and files enriched with new objects or characters;
(3) INTERROGATION, during which information is sought according to defined criteria, and extracted as required (Fig. 3);
(4) REVISION, during which structural information may be redefined, equivalent to a restructuring of the data file.

The system is extremely easy to use, particularly for extracting and printing information, but is incapable of running large databases because of its sequential organization. It is, therefore, clear that INFOL is unsuitable as the final language for our database; meanwhile, it can be used for data management in the various projects mentioned above. The structure of INFOL and its flexible use of files convinced us to score data as "free text", attempting to reduce the use of codes to a minimum. This step allows us to avoid returning to the raw data when the final database is created. It was also better to divide information into several categories:

(1) nomenclatural and systematics data: we thought it of primary importance to be able to trace the evolution of ideas concerning each taxon, by recording all synonyms and determinavits;
(2) geographical descriptors, including the district and precise locality using coordinates;
(3) descriptors concerning the plant's environment, soil substratum, ecology and population parameters;
(4) description of the plant itself and its uses.

Whilst wishing to preserve a certain unity during data collection in the field or in the herbarium, it was difficult to design a single form equally valid for European and overseas crops. Initially we therefore created, together with the project heads, a taxonomic data sheet including all the information we wished to register and a habitat data sheet for local, European and tropical flora. For each of these, a simplified sheet was prepared, allowing direct floristic data capture in the field. On the reverse of the taxonomic forms, we left a number of free spaces for registering morphological, cytological and bio-

```
ESTABLISHMENT
ITEM DESCRIPTION 25
Numéro d'acquis.  * 1* Nom après déter.  * 2*
Determinavit      * 3* Date              * 4*
Nom reçu          * 5* Réf. culture      * 6*
Famille           * 7* Synonymes         * 8*
Origine           * 9* Provenance        *10*
Type reçu         *11* Ecologie          *12*
Date semis        *13* Mélange terreux   *14*
Empl. conseillé   *15* Date plantation   *16*
Empl. définitif   *17* Herbier           *18*
Récolte graines   *19* Photo             *20*
A Eliminer        *21* Disparitions      *22*
Nom commun        *23* Référence         *24*
Observations      *25*

CATEGORY-TYPE
* 1* UNAR NUME * 2* UNAR ALPH * 3* MULT ALPH
* 4* UNAR DATE * 5* UNAR ALPH * 6* MULT CODE
* 7* MULT ALPH * 8* MULT ALPH * 9* UNAR ALPH
*10* UNAR ALPH *11* UNAR ALPH *12* UNAR ALPH
*13* UNAR DATE *14* UNAR ALPH *15* MULT ALPH
*16* UNAR DATE *17* UNAR ALPH *18* UNAR CODE
*19* UNAR CODE *20* UNAR CODE *21* UNAR CODE
*22* UNAR ALPH *23* UNAR ALPH *23* UNAR ALPH
*24* UNAR ALPH *25* UNAR ALPH

CODES
* 6* MNEMONIC
BJ   Bon Jardinier *
BT   Binz & Thommen: Flore de Suisse *
FE1  Flora Europea T. 1 * FE2 Flora Europea T. 2 *
FE3  Flora Europea T. 3 * FE4 Flora Europea T. 4 *
FE5  Flora Europea T. 5 *
FF   Fournier: Quatre Flores de France *
K    Index Kewensis *
Z    Zänder (Handwörterbuch der Pflanzennamen) *
*18* MNEMONIC   X  Oui *
*19* CODED  AS *18*
*20* CODED  AS *18*
*21* CODED  AS *18*   *

VALIDATIONS
* 3* MAXIMUM 2 * 7* MAXIMUM 2
* 8* MAXIMUM 3 *10* NECESSARY
```

Fig. 2. An ESTABLISHMENT description (Catalogue of the Botanical Garden).

chemical data for the use of each research worker, according to his needs (Mascherpa, 1976; Aeschimann *et al.*, 1981; Jeanmonod and Mascherpa, 1982).

At present the information scored is divided among a series of files between which there is no direct link. In each case the links are made using special programs. An undergraduate project has just been started, in conjunction with the UCC, to study methods and the structure of botanical information. Several other projects have begun

```
INTERROGATION
RETRIEVAL CRITERIA 1
*1*  ( GE 820103 AND LE 820196 ) OR ( GE 820589 AND LE 820824 )
AND ( *  7* EQ *Asteraceae* OR EQ *Compositae* )

EXTRACTION 1
DESCRIPTION PAGEWIDTH 79 ELEMENT LENGTH 63 ACCEPT FORMAT
*  1* ROW  1 COLUMN  1
*  2* ROW  1 COLUMN 10
*  5* ROW  3 COLUMN  1 WITH DESC
*  3* ROW  3 COLUMN 60 WITH TEXT Dét. : *
*  4* ROW  4 COLUMN 60 WITH TEXT Date : *
*  7* ROW  5 COLUMN  1 WITH DESC
*  6* ROW  5 COLUMN 60 WITH TEXT Réf. : *
*  8* ROW  6 COLUMN  1 WITH DESC
*  9* ROW  8 COLUMN  1 WITH DESC
*11* ROW  8 COLUMN 60 WITH TEXT Type: *
*10* ROW 10 COLUMN  1 WITH DESC
*12* ROW 16 COLUMN  1 WITH DESC
*15* ROW 16 COLUMN 60 WITH TEXT Empl.: *
*14* ROW 20 COLUMN 21 WITH TEXT Mélange: *
*13* ROW 21 COLUMN  1 WITH TEXT Semis: *
*16* ROW 22 COLUMN  1 WITH TEXT Plant : *
*17* ROW 24 COLUMN  1 WITH TEXT Empl.  définitif: *
*18* ROW 24 COLUMN 60 WITH TEXT Herbier: *
*19* ROW 25 COLUMN 60 WITH TEXT Graines: *
*20* ROW 26 COLUMN 60 WITH TEXT Photos : *
*21* ROW 27 COLUNN 60 WITH TEXT Eliminer: *
*22* ROW 28 COLUMN  1 WITH DESC
*  0* ROW 63 COLUMN  1 WITH TEXT --------------------------- *
*23* ROW 34 COLUMN  1 WITH DESC
*24* ROW 36 COLUMN  1 WITH DESC
*25* ROW 48 COLUMN  1 WITH DESC
*  0* ROW 62 COLUMN  1 WITH TEXT Conservatoire et Jardin botaniques *
*  0* ROW 63 COLUMN  1 WITH TEXT Genève--------------------- *
FINIS FINIS
```

Fig. 3. An INTERROGATION description (Catalogue of the Botanical Garden).

to fill in information both for the herbarium and library for the *Cyperaceae, Poaceae, Araceae, Commelinaceae, Juncaceae, Chenopodiaceae, Polygonaceae, Fabaceae, Ranunculaceae* and *Orchidaceae.* Only once these projects have been finished and the results gathered, shall we be able to suggest a final scheme for our database.

2. Additional Programs

In addition to programs for managing addresses and files, for preparing computer listings for automatic editing, etc., we have introduced two complementary programs – one for mapping and the other for automatic identification.

(a) Mapping. To avoid distribution lists which are commonly produced from databases, we felt that automatic preparation of distribution maps was desirable. To accomplish this, two problems needed to be solved:
(1) digitalization of maps and their reproduction on computer hardware;
(2) linking with the databases.
In INFOL, links to a sequential file are easily arranged using special subroutines to extract selected data. The automatic production of maps, however, requires the coordinates of the selected taxa, as well as the digitalization of the maps themselves, and is a far more interesting problem.

Since at first we were unable to create a larger number of maps, we decided to digitalize each individually, on request. They are held in a file addressed directly by the database package. We are thus obliged to digitalize every published map, despite not knowing either the type or the projection used. We developed a program for smoothing curved coordinates which requires a reference grid standard to be recorded for each map. Although highly flexible (one may work on any data set) the program requires a fairly lengthy calculation, and will therefore be abandoned at a later stage. We shall probably have to orthogonalize coordinate grids and reduce the number of types of map offered.

(b) Identification. Automatic identification has greatly interested us since the discovery of Pankhurst's volume (1975), particularly his interactive program. Early tests were made on the UNIVAC and a

VIDEOPLAN microcomputer of 64K bytes (Mascherpa and Bocquet, 1981). If automatic identification programs are to be used for regional flora in a long-term project it is necessary to provide an interactive package for the management of a taxon/character matrix, permitting the capture and correction of data, the definition of new characters or the suppression of characters already scored (Hubert van Blijenburgh, 1982).

The program written was aimed at maximum use of the interactive capabilities of a large computer, whilst keeping only the minimum of information in central memory; we have relied on direct access from high-capacity discs. The description of characters and their states and the values attributed to each diagnostic character in the matrix, are manipulated using a series of pointers which appear at the start of the file. These pointers are accessed directly at the initiation of the program. This flexible way of recording data is also advantageous during data capture at the terminal, since one may divide the matrix into blocks and halt data entry whenever required. Automatic identification of the Swiss Flora is based on diagnostic characters grouped by theme, using a two-level menu. Data is recorded in a direct-access file and our techniques follow those of Pankhurst and Aitchison (1975). Emphasis has been placed on the definition of botanical terms and use of an interactive "user-friendly" program.

CONCLUSIONS

The introduction of new techniques, like computerization, into a nearly two-hundred-year-old museum, is not easy. Alongside tradition and customary practice one must open new routes, trying, above all, to propose original solutions and help with the work carried out in the large herbaria, while at the same time avoiding too high an increase in the workload of technicians and research workers responsible for the collections. There are no ready-made solutions to these problems. The logical course we have followed is to decide at each step what is really the botanist's or technician's task, what are the problems and which solutions are usually adopted.

We have attempted to develop techniques for collating botanical information that reconstruct the botanist's approach. Only later, when the various projects come to their conclusion, will we suggest a system for constructing a database for our Institution. It is obvious that, at the present stage in the project, we have wished to call upon student theses to avoid overloading the Institute's permanent staff. The future database, however, could not be built without the help of new personnel, materials and grants. Finally, from the programs produced to manage the database, it appears even clearer to us that writing programs to manage the information is a small task compared with the enormous effort required to improve communications between botanist and computer. Nevertheless, only a few people are interested in the problems of dialogue between human and machine.

REFERENCES

Aeschimann, D., Mascherpa, J. M. and Bocquet, G. (1981). Etude biosystématique du *Silene vulgaris* s.l. (Caryophyllaceae) dans le domaine alpin. Méthodologie. *Candollea* **36**, 379–396.

Bertoni, B. S., Mascherpa, J. M. and Spichiger, R. (1982). Datos bibliográficos para el estudio de la vegetación y de la flora del Paraguay. *Candollea* **37**, 277–313.

Bocquet, G. (1982). Activités des Conservatoire et Jardin botaniques en 1981. *Candollea* **37**, v–xiii.

Bocquet, G. and Mascherpa, J. M. (1982). Informatique: réalisations et projets. *Candollea* **37**, xxxvii–xiii.

Brenan, J. P. M., Ross, R. and Williams, T. (1975). "Computers in Botanical Collections". Plenum Press, New York.

Crovello, T. J. (1972). Computerization of specimen data from the Edward Lee Greene Herbarium (ND-G) at Notre Dame. *Brittonia* **24**, 131–141.

Crovello, T. J. and Macdonald, R. D. (1970). Index of EDP-IR projects in systematics. *Taxon* **19**, 63–76.

Encarnación, F., Spichiger, R. and Mascherpa, J. M. (1982). Bibliografía selectiva de las familias y de los géneros de Fanerógamas. Primera contribución al estudio de la flora y de la vegetación de la Amazonia peruana. *Boissera* **34**, 1–197.

Forero, E. and Pereira, F. J. (1976). EDP-IR in the National Herbarium of Columbia (COL). *Taxon* **15**, 85–94.

Gavin, P., Agopian, P., Keller, L.-D., Perret and Villard, H. (1980). "SIBIL. Système intégré pour les bibliothèques universitaires de Lausanne". 2nd edition. Bibliothèque Cantonale et Universitaire de Lausanne.

Gómez-Pompa, A. and Nevling, L. I. (1973). The use of electronic data processing methods in the flora of Veracruz program. *Contr. Gray Herb.* **203**, 49–64.

Greuter, W., Burdet, H. M. and Long, G. (eds) (1981). "MED-CHECKLIST. I. Pteridophyta". OPTIMA, Geneva and Berlin.

Hall, A. V. (1974). Museum specimen record data storage and retrieval. *Taxon* **23**, 23–28.

Hubert van Blijenburgh, N. W. P. (1982). "Détermination automatique des familles de la Flore de Suisse: Formation de la matrice des caractères, écriture du programme de détermination". Travail de diplôme, Université de Genève.

Hunt, D. R. (1977). "The living collections records file". Internal report, Royal Botanic Gardens, Kew.

Iff, J. (1982). Jardin. *Candollea* 37, xiii–xvii.

Jeanmonod, D. and Mascherpa, J. M. (1982). Révision de la section Siphonomorpha Otth du genre *Silene* L. (Caryophyllaceae) en Méditerranée occidentale. Méthodologie. *Candollea* 37, 497–524.

Levrat, B., Baud, J.-P., Jacquesson, A., Chenais, J.-M. and Levêque, F. (1979). "INFOL-2. Manuel de référence". 3rd edition. Centre Universitaire d'Informatique, Université de Genève.

Mascherpa, J. M. (1976). "Application des méthodes informatiques à la taxonomie végétale, notamment des Phaseolinae, et aux problèmes biologiques qui lui sont liés". Thèse no. 1767, Université de Genève [Coopérative d'imprimerie du Pré-Jerôme], Genève.

Mascherpa, J. M. and Bocquet, G. (1981). Deux programmes interactifs de détermination automatique. Une idée, un but. *Candollea* 36, 463–483.

Morris, J. W. and Manders, R. (1981). Information available within the PRECIS data bank of the National Herbarium, Pretoria, with examples of uses to which it may be put. *Bothalia* 13, 473–485.

Pankhurst, R. J. (ed.) (1975). "Biological Identification with Computers". Academic Press, London and Orlando.

Pankhurst, R. J. and Aitchison, R. R. (1975). A computer program to construct polyclaves. *In* "Biological Identification with Computers" (R. J. Pankhurst, ed.), pp. 73–79. Academic Press, London and Orlando.

Spichiger, R. (1982). Recherches tropicales. *Candollea* 37, xx.

Sweet, H. C. and Poppleton, J. E. (1977). An EDP technique designed for the study of a local flora. *Taxon* 26, 181–190.

21 | Identification of Toxic Mushrooms and Toadstools (Agarics) — An On-line Identification Program

P. MARGOT

Federal Institute of Technology,
Lausanne, Switzerland

G. FARQUHAR

Computer Centre, Strathclyde University,
Glasgow, UK

and

R. WATLING

Royal Botanic Garden, Edinburgh, UK

Abstract: The characters used in the identification of toxic mushrooms and toadstools are discussed and the ways in which they are presented for an on-line identification program given. Examples are offered by way of illustration.

INTRODUCTION

Many more people are experimenting with eating natural foods, including mushrooms and toadstools, than ever before, and hallucinogenic fungi are increasingly being used as "recreational" drugs in the

Systematics Association Special Volume No. 26, "Databases in Systematics", edited by R. Allkin and F. A. Bisby, 1984, pp. 249–261. Academic Press, London and Orlando.

DATABASES IN SYSTEMATICS
ISBN 0 12 053040 6

British Isles. Although intoxication in the past has been relatively rare there has been a worrying increase in the number of cases brought to the attention of the medical profession in the last few years.

MUSHROOM IDENTIFICATION: CHARACTER USEFULNESS

It was considered that a means by which fragments of agaric fruit-bodies (basidiomes) could be easily identified was long overdue (Margot *et al.*, 1980). A rapid and accurate service was imperative as the majority of British professionals becoming more involved, i.e. doctors, toxicologists and mycologists, are not suitably conversant with emergencies involving fungi.

The traditional approach to the identification of the agarics by marrying the results from the examination of macro- and microscopic characters is inappropriate for dealing with material taken from patients' meals, persons or homes, or with confiscated material from miscreants. The use of macroscopic characters has dominated agaricology for over 150 years and only comparatively recently has the use of the microscope been accepted worldwide as a means of identifying species and arranging those species into families.

Agaric families are defined on a few fundamental micro-characters which are easily observed even in fragmented material where macroscopic characters are lacking. The techniques by which these structures are observed have been used over a 15-year period by Watling to teach a wide spectrum of people, from interested amateurs to university students and professionals.

Agarics can be divided into those in which the flesh (trama) of the cap is composed of a mixture of packets of round cells and long cylindric, flexuous hyphae (heteromerous: Fig. 1D and 1E) and those with a flesh homogeneous in its construction (homoiomerous: Fig. 1F and 1G). In homoiomerous species the flesh between the two adjacent gill-faces (hymenophoral trama) might contain hyphae that are bilateral, convergent, regular or irregular (Watling, 1973). The outermost layer of the cap (or pileus) of the basidiome may be composed of rounded cells (Fig. 1Q) or filamentous cells (Fig. 1R); a scalp of the pileus is then known as "cellular" or "filamentous". This microscopic data offers powerful information when correlated with the colour of the spore-deposit, often only ascertainable by examination of the basidiospores in water or dilute aqueous solutions of alkali. All these are very positive characters and lend themselves to manipulation for computer analysis.

Thus the Fly agaric, *Amanita muscaria*, a popular recreational mushroom has homoiomerous pileus trama, bilateral hymenophoral trama, white spore-print and a "filamentous" cuticle in contrast with the non-psychotomimetic, non-toxic *Lactarius tabidus* which possesses a white spore-print, "cellular" cuticle and heteromerous cap and hymenophoral tramas. Therefore, with only a few characters, several major groupings can be obtained. The techniques by which these characters can be demonstrated have been outlined by Watling (1968 and 1983). This microscopic data should take no more than 5 min to obtain whereas some generic and certainly species differentiation is more difficult, unless further microscopic characters are utilized and these now are routinely used by most mycologists. They include the dimensions and shape of the basidiospores (Fig. 2; Table 1) and the presence or absence of specialised sterile cells in the hymenium termed cystidia, and when present the morphology of these structures (Fig. 1H–P; Table 2). Finally, the presence or absence of clamp-connections (Fig. 1G) and a number of spores per basidium are useful characters to record, e.g. the N. American *Amanita bisporigera* differs from *A. virosa*, also white, by possessing 2-spored basidia (Fig. 2L and M).

Table 1. Characters and states of basidiospores used for data collection.

Colour	hyaline, brown, honey-coloured or pinkish, dark (purple-brown, black etc.).
Shape	globose – subglobose (Fig. 2Q), ellipsoid (Fig. 2L), angular (Fig 2B, 2J), otherwise (e.g. Fig. 2G).
Wall thickness	thin (Fig. 2A), thick (Fig. 2N).
Surface	smooth (Fig. 2P), rough (Figs 2C, 2F).
Germ-pore	present (Fig. 2E), absent (Fig. 2J).
Length (μm)	<6 (Fig. 2K), 6–8 (Fig. 2G), 8–10 (Fig. 2B), 10–12 (Fig. 2H), 12–17 (Fig. 2E), >17 (Fig. 2N).
Breadth (μm)	<4 (Fig. 2G), 4–6 (Fig. 2I), 6–8 (Fig. 2M), 8–10 (Fig. 2L), >10 (Fig. 2E).
Colour of spores in mass	white, pink, brown, black.

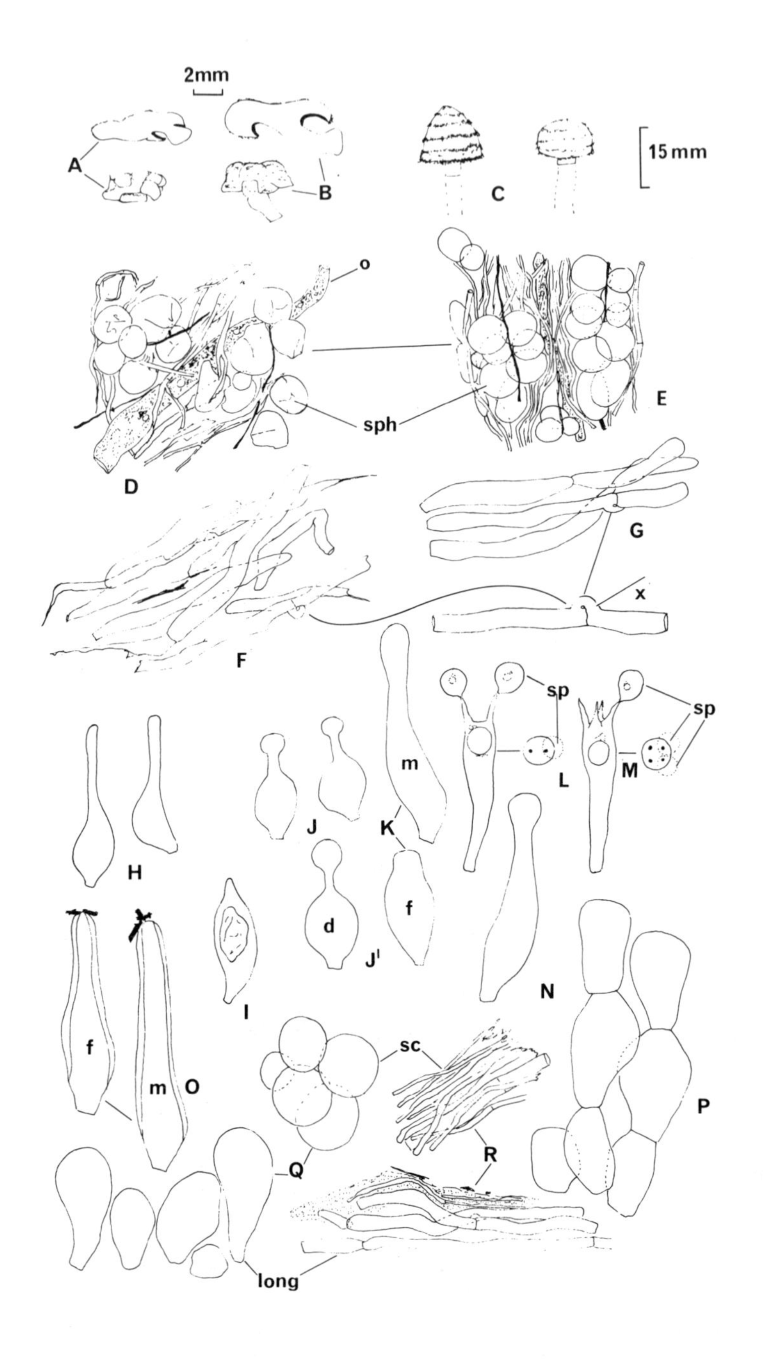
2mm
15 mm
A
B
C
o
sph
D
E
G
x
F
sp
sp
m
H
J
K
L
M
f
d
J'
I
N
f
m
O
sc
P
Q
R
long

Table 2. Characters and states of cystidia used for data collection.

Marginal cystidia (μm)	absent (Fig. 1P), small (<35), medium (35–55), large (>55).
Marginal cystidia shape	ampulliform (Fig. 1H), lageniform, lecythiform (Fig. 1J), utriform, metuloid (Fig. 1O).
Facial cystidia (μm)	absent, small (<35), medium (35–55), large (>55).
Facial cystidia shape	ampulliform, lageniform, lecythiform, utriform (Fig. 1K), metuloid (Fig. 1O).
Chrysocystidia	present (Fig. 1I), absent.

Fig. 1. A, bean-shaped objects confiscated from miscreant, one sectioned; B, same partially imbibed with water; C, same treated with 10% aqueous solution of ammonia to show origin; D, flesh from basidiome involved in poisoning of horse; E, flesh of Russulaceae heteromerous trama with oleiferous hyphae (o) and sphaerocysts (sph); F, flesh from dried *Panaeolus subbalteatus*; G, same from fresh material: homoiomerous trama; one hypha isolated to show clamp-connections (x); H, marginal cystidia in Magic mushroom (*Psilocybe semilanceata*); ampulliform cheilocystidia; I, Chrysocystidium in *Pholiota squarrosa*, a taxon said to be toxic; J, marginal cystidia in *Conocybe semiglobata*; lecythiform cheilocystidia; J^1, Dermatocystidium (d) from stipe; K, marginal (m) and facial (f) cystidia in Mexican Gold Tops (*Psilocybe cubensis*); utriform pleurocystidium; L, basidium of *Amanita bisporigera*, a N. American poisonous fungus; also shown end-on; M, basidium of Destroying Angel (*A. virosa*); also shown end-on. sp. = basidiospore; N, marginal cystidium in *Panaeolus subbalteatus*; subcapitate cheilocystidium; O, marginal (m) and facial (f) cystidia in *Inocybe geophylla*; metuloids; P, marginal cells in chains in *Amanita muscaria*; Q, cells of cap "cuticle" (pileipellis) of *Panaeolus sphinctrinus* in longitudinal section (long and in scalp (sc); R, as for Q, *Psilocybe semilanceata*.

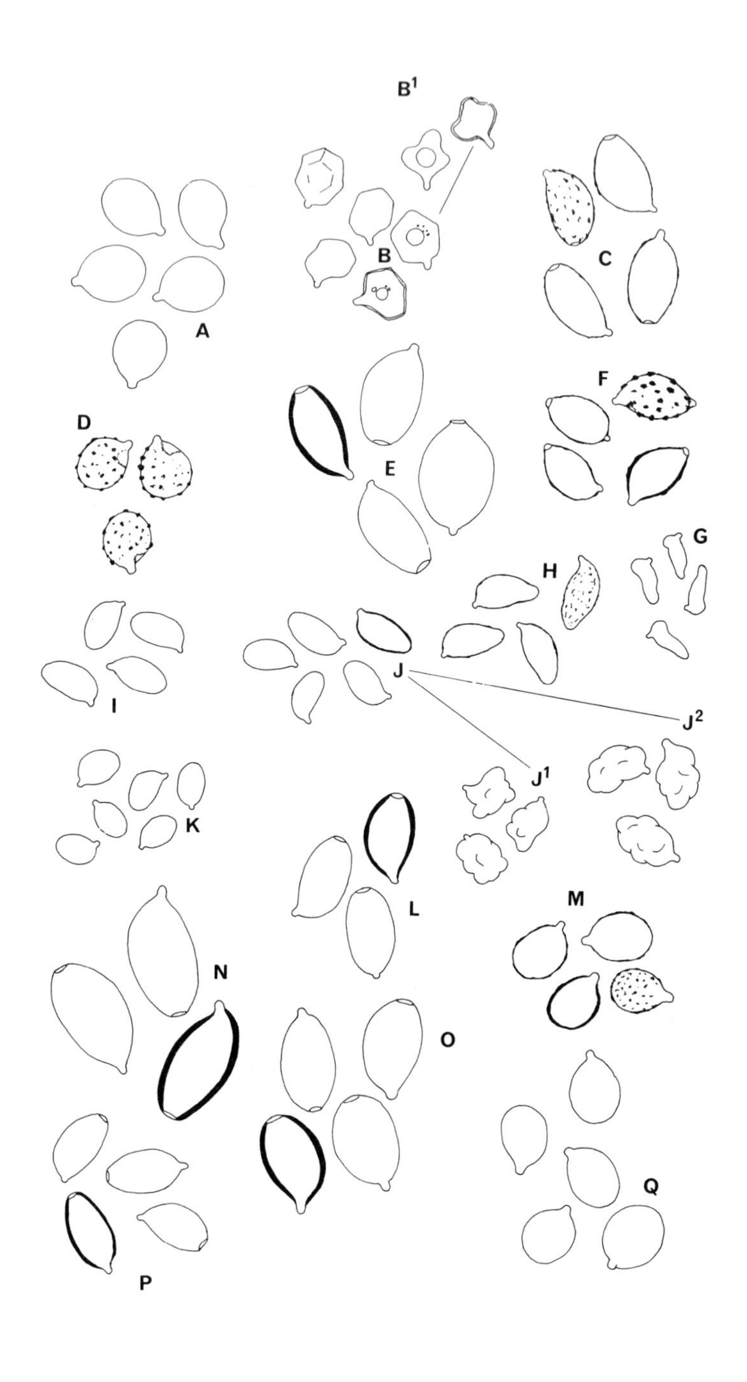
B1
B
A
C
D
E
F
G
H
I
J
J2
J1
K
L
M
N
O
P
Q

Whilst it is necessary to resort to macroscopic characters such as pileus (cap) and stipe (stem) colour, texture and shape for a definitive identification, luckily the small amount of data described above (24 characters) is sufficient to recognize groups of clearly related species and it is frequently the case that related species possess the same toxin, taxa differing one from the other in magnitude or quantity.

To the information available from the actual specimens, three ecological classes (16 states in all) have been added. This ecological data can only be obtained from the patient, inference, etc. and is therefore open to doubt or interpretation. Any medical data and chemistry involving the toxins is not open to doubt but as in so many cases symptoms must be carefully ascertained. Generalizations can be made and even a simple flow-diagram has been found to be of great use in suggesting the cause of poisoning (Pegler and Watling, 1982).

CHARACTER PRESENTATION

The characters outlined above have not been weighted for their usefulness in separating taxa for anything other than their ease of observation. Thus a multi-access key can be constructed which, although purely artificial, reflects, in general terms, the present knowledge not only of poisonous mushrooms and toadstools but also the entire agaric classification.

Margot (1980) has collated the characters outlined above for 126

Fig. 2. Basidiospores isolated in actual cases of poisonings. A, *Amanita muscaria*; B, *Nolanea sericea* (B[1] *N. staurospora*); C, *Panaeolina foenisecii*; D, *Russula olivacea* showing amyloid ornamentation; E, *Panaeolus sphinctrinus*; F, *Lacrymaria glareosa*; G, *Lepiota cristata* – spores éperonées; H, *Hebeloma ? crustuliniforme*; I, *Paxillus involutus*; J, *Inocybe geophylla* var *lilacina*. (J[1]. *I. napipes*; J[2], *I. praetervisa*); K, *Agaricus xanthodermus*; L, *Conocybe semiglobata*; M, *Cortinarius speciosissimus*; N, *Psilocybe cubensis*; O, *Panaeolus subbalteatus*; P, *Psilocybe semilanceata*; Q, *Amanita phalloides* – from dried material. Reference should be made to descriptive data in Table 1. Names in parenthesis are of species suspected of being toxic but not found by Watling in poisonings.

species of agarics of suggested toxicity. He includes references to three publications for each species, one describing the taxon, the second describing the type of toxin involved and the third describing the medical aspects and the treatment in case of emergency. A second stage increased the number of species to over 200 suspected toxic agarics suggested as dangerous by claims in the literature or by chemical analysis and recognition of the toxin therein (Greer, 1980; Watling, 1982).

Margot decided how to deal with all morphological data pertaining to spores and cystidia, and quite artificial groups were produced (see Table 1). Nevertheless, they were found to be practical and the whole system was tested on a number of occasions (about 50 identifications, composed of 30 actual cases and 20 runs by an expert mycologist).

In any one sample it may not be possible to ascertain all the micro-characters and certainly the content or identity of the toxin are not yet known for many. This makes the program proposed by Margot suspect, except that a trip-system was incorporated to give a "first-miss" identification. This has certainly proved useful and the method is described below.

ON-LINE PROGRAM

The basic idea of the program is that given information supplied by the user, a search of a database of known species is carried out and the user informed of any matching species found. The program is intended as a tool for the non-computer specialist and as such it was designed to be simple in appearance and easy to use. A menu-driven approach has been adopted in which the user is simply presented with numbered lists of items from which to make relevant selections. For a general review of biological identification using computers see Morse (1975).

The original implementation of the program was in Fortran on the local Honeywell 66/40 at Strathclyde University. Versions in Applesoft Basic and Microsoft Basic also exist. The Fortran implementation uses a subset of the language, which is as machine independent as possible to aid portability. The program occupies 35K bytes when loaded, and uses a file of approx. 140K bytes of information.

The database used consists of information on 224 species of

poisonous mushroom. The information on each species is essentially a mixture derived from observations of the characters of various species both in the field and in the laboratory, together with text information such as species name, toxin present, medical treatment and bibliographic references. Twenty-four characters have been included in the program with 117 possible states.

Several of the characters, e.g. toxins and clamp-connection have mutually exclusive discrete states which may occur in particular species; whereas other characters, such as spore length and spore breadth, because of their essentially continuous nature, may have state overlaps. These last two cases initially posed problems in data representation. Each state within a character is assigned an integer value of one less than the number in the corresponding menu list seen by the user. Mutually exclusive states are simply denoted in the database by their state number whereas the simple technique of replacing variable characters by the construction "−1, count, state 1, state 2, etc." alleviates the problem of multistate representation. The negative integer indicates to the program that a variable character is about to be encountered. The fields following provide a count of the states present and the actual state values themselves.

By adopting this simple compression technique the values to be stored in the database for a given species have been reduced from 78 to around 45. A typical example of a database entry is shown in Table 3. Each database entry is expanded to its full 78 values by introducing appropriate redundant items, before any attempt at comparison with the input data is made. This is simply a matter of

Table 3. Typical database entry.

5	0	2	3	−1	2	1	2	3	0
1	0	1	0	2	2	3	3	2	0
0	1	0	−1	2	0	1	0	0	0
−1	2	1	2	0	0	−2	2	1	1

PSILOCYBE SUBAERUGINASCENS HOHN PSILOCYBIN/INDOL HALL PROPOSED TREATMENT PREVENT ABSORPTION, ENHANCE EXCRET, RESTRAIN SELF DESTRUCT BEHAV, QUIET, POS DIAZEPAM, CHLORPROMAZINE TAXON SINGER & SMITH (1958) MYCOLOGIA 50, 271-27 CHEM HOFMANN ET AL (1959) HELV CHIM ACTA 42,1557-1572 MED LINCOFF & MTC (1977) TOX & HALLUC. MUSHROOM POISONING, VAN NOSTRAND REINHOLD, NY. HALLUCINOGEN, UPSETTING E. & S.E. ASIA.

computational convenience: it is easier to examine fixed length data than variable length data.

As indicated earlier the program is menu driven: no complicated questions are asked. Nearly all user input is in numeric form, the only exception being a simple yes/no response required after a matching species has been found, in order that the program continues its search. Based on the information supplied by the user, a user-search vector, is constructed. The form of the user-search vector is identical to that of the fully expanded numeric part of the database entry. It is this search vector that is used to search the database for appropriate species.

For certain characters such as spore length, spore breadth and time of growth, because of their essentially continuous nature the wrong state may be chosen and the program will then attempt, if the user wishes, to identify "near misses". This means, for example, that although a particular specimen under examination has been identified, say as a summer species, the program can be instructed to also treat it as if it had been recognized as a spring or autumn species. This will provide a wider spread of data with which to attempt a match.

In order to make database maintenance as easy as possible the most appropriate file structure available to support the data was that of the simple serial file. Random access files, although probably better for data chaining and searching techniques, lead to problems in data maintenance. The software tools required to maintain and extend a direct access database need to be more sophisticated than those applicable to serial files, where appropriate text editors may be used.

The example shown below of an exact match using only three characters indicates the user's view of the program (user responses are preceded by an "=" sign). Only two of the species found by the program have been included (the program actually located eight possible species).

Clearly, the results produced by the program depend on the information supplied by the user. The program cannot replace the expert mycologist where a positive identification between two closely related species of mushroom is required. It is intended to fill a gap in the rapid treatment and identification of mushroom poisoning and to provide a tool which can be used to reduce considerably the number of possibilities.

EXAMPLE OF A TYPICAL PROGRAM RUN

FUNGI IDENTIFICATION PROGRAM

1. SYMPTOMS
2. ONSET OF SYMPTOMS
3. GEOG. DISTRIBUTION
4. HABITAT
5. TIME OF GROWTH
6. SPORE COLOUR UNDER M.SCOPE
7. SPORE SHAPE
8. SPORE SURFACE
9. SPORE WALL THICKNESS
10. GERM PORE
11. SPORE LENGTH
12. SPORE BREADTH
13. SPORE PRINT
14. TOXINS
15. DURATION OF SYMPTOMS
16. PILEUS TRAMA
17. GILL TRAMA
18. CUTICLE
19. CLAMP CONNECTION
20. BASIDIA/ASCI NO. OF SPORES
21. BASIDIA/ASCI SIZES
22. MARGINAL CYSTIDIA
23. FACIAL CYSTIDIA
24. KOH CHRYSOCYSTIDIA

WHICH CHARACTERS DO YOU HAVE DATA FOR?
ENTER CHARACTER NUMBERS COMMA SEPARATED
=3, 4, 7

DATA FOR 3 CHARACTERS REQUIRED

IF NO SPECIES MATCH EXACTLY THE GIVEN DATA A CLOSE MATCH CAN BE ATTEMPTED ON THE FOLLOWING CHARACTERS:-

4. HABITAT
5. TIME OF GROWTH
9. SPORE WALL THICKNESS
11. SPORE LENGTH
12. SPORE BREADTH

PLEASE GIVE THE CHARACTER NUMBERS YOU REQUIRE (COMMA SEPARATED) OR 0(ZERO) IF NO CLOSE MATCHING IS TO BE ATTEMPTED
=0
NO CLOSE MATCHING REQUIRED

3. GEOGRAPHICAL DISTRIBUTION
   1. N. TEMPERATE
   2. S. TEMPERATE
   3. SUBTROPICAL

```
        4. TROPICAL
        5. ENGLISH
        6. SCOTTISH
OPTION NUMBERS?
ENTER DATA COMMA SEPARATED
=1,6
OPTIONS ACCEPTED --
        1. N. TEMPERATE
        6. SCOTTISH

4. HABITAT
        1. GROUND IN CONIF. WOODS
        2. GROUND IN HARDWOODS
        3, GROUND IN GRASS
        4. GROUND IN DUNG
        5. ON STUMPS
        6. OTHERS
OPTION NUMBERS?
ENTER DATA COMMA SEPARATED
=1
OPTIONS ACCEPTED –
        1. GROUND IN CONIF. WOODS

7. SPORE SHAPE
        1. GLOBOSE
        2. ELLIPSOID
        3. ANGULAR
        4. VARIOUS
OPTION NUMBERS?
=4
OPTION 4 -- VARIOUS

PROGRAM RUNNING
POSSIBLE SPECIES (EXACT MATCH ON SUPPLIED DATA)-
INOCYBE ALBODISCA PECK UK FREQ. 0.4 MUSCARINE
PROPOSED TREATMENT:PREVENT ABSORP, ENHANCE
EXCRET, ATROPINE UNTIL CESSATION SECRETIONS,
SUPPORTIVE TAXON PECK (1898) ANN REP NY ST. MUS
51,290 CHEM CATALFOMO & EUGSTER (1970) HELV CHIM
ACTA 53:4,848-851 MED LINCOFF & MITCHEL (1977) TOX
```

& HA MUSHROOM POISONING, VAN NOSTRAND REINHOLD, NY..
UPSETTING TO DANGEROUS S.EU : N. AM

POSSIBLE SPECIES (EXACT MATCH ON SUPPLIED DATA) –
INOCYBE FIBROSA (SOW) BRES MUSCARINE PROPOSED
TREATMENT; PREVENT ABSORPTION, ENHANCE EXCRET,
ATROPINE UNTIL CESSATION SECRETIONS, SUPPORTIVE
TAXON HEIM (1931) LE GENRE INOCYBE, LECHEVALIER,
PARIS CHEM CATALFOMO & EUGSTER (1970) HELV CHIM
ACTA 53:4,848-851 MED LINCOFF & MITCHEL (1977) TOX
& HALLUC MUSHROOM POISONING, VAN NOSTRAND
REINHOLD, NY
UPSETTING TO DANGEROUS EU; N. AM
END OF PROGRAM

REFERENCES

Greer, F. A. (1980). Chemical investigation on the toxins of *Cortinarius speciosissimus* and related mushrooms. M.Sc. thesis, Strathclyde University, Glasgow.

Margot, P. (1980). On-line identification program of poisonous and hallucinogenic mushrooms. *In* "Forensic Toxicology" (J. Oliver, ed.), pp. 221–234. Croom Helm, London.

Margot, P., Farquhar, G. and Tilstone, W. J. (1980). Application to macro- and micro-computers of an identification programme for poisonous and hallucinogenic mushrooms of forensic interest. *Proc. First Scand. Meeting For. Sci. Linköping*, pp. 145–152.

Morse, L. E. (1975). Recent advances in the theory and practice of biological specimen identification. *In* "Biological Identification with Computers" (R. J. Pankhurst, ed.), pp. 11–52. Academic Press, London and Orlando.

Pegler, D. N. and Watling, R. (1982). British toxic fungi. *Bull Br. Mycol. Soc.* **16**, 66–75.

Watling, R. (1968). Hints on the microscopical examination of the Agaric fruit-body. *Microscopy* **31**, 95–105.

Watling, R. (1973). "Identification of the Larger Fungi". Hulton, Amersham.

Watling, R. (1982). *Cortinarius speciosissimus*: The cause of renal failure in two young men. *Mycopathologia* **79**, 71–78.

Watling, R. (1983). "Hallucinogenic Mushrooms". *J. Forensic Sci. Soc.* **23**, 53–66.

22 Handling Taxonomic Descriptions by Computer

R. ALLKIN

*Biology Department, The University,
Southampton SO9 5NH, UK*

Abstract: To build taxonomic information systems we need to represent taxonomic descriptions in a format suitable for computer storage and yet flexible enough for a variety of needs. This is not straightforward. For information retrieval purposes all elements of descriptions should be strictly comparable. This requires that terminology be used consistently and that, homology permitting, precisely the same features be described for all taxa. The latter is rarely achieved when specialist terminology is used. Taxonomists employ a wide range of descriptor types and a flexible data format is therefore essential. Personal opinions and frequency estimates may also be recorded but cannot be incorporated in a simple rectangular data structure. Use of the taxonomic hierarchy and the logical relationships among characters can facilitate data verification and assist efficient data capture, storage and retrieval. Unfortunately, inclusion of these relationships adds further complexity to the data structure. Intrataxon variability requires provision both for more than one value per data cell and for the description of "complexed" variation patterns. The multiple use intended for descriptive data causes conflicts during character definition and selection. The inclusion of "sister" characters and of "apparent" descriptions is discussed.

INTRODUCTION

A number of current projects aim to build databases containing that information conventionally printed as monographs, Floras or Faunas. Such projects need to store taxonomic descriptions and have access to a data-management system capable of manipulating them as

Systematics Association Special Volume No. 26, "Databases in Systematics", edited by R. Allkin and F. A. Bisby, 1984, pp. 263–278. Academic Press, London and Orlando.

DATABASES IN SYSTEMATICS
ISBN 0 12 053040 6

required. This paper is concerned principally with the first of these requirements: the definition of a data format suitable for taxonomic descriptions.

Taxonomists routinely handle large collections of descriptive biological data. We store, revise and arrange this data, analyse it and make critical assessment. Finally, we should communicate both the results of analyses and, equally important, the raw data used (obtained either by observation or from the literature). Our dream (Bisby, 1984) is that descriptive taxonomic data, stored in computer files will be manipulated using computer programs to accomplish all of the necessary tasks (see Fig. 3, Bisby, Chapter 2, this volume). It is argued that data manipulation will be more flexible, data retrieval far more powerful and data capture more efficient: one data set serving many purposes (Allkin and White, 1982).

Flexible data management and information retrieval are possible only if descriptions are stored in a structured manner. Textual descriptions stored in computer files have much more limited value. Whether or not taxonomists encode their data during data capture is not relevant here. What is important is that explicitly defined characters are used with discrete sets of alternative states. Storing descriptions, at least as we are accustomed to using them, in such a structured manner is unfortunately not straightforward. To this audience I will play the devil's advocate and examine properties of taxonomic descriptions which cause difficulty. My observations are drawn largely from my experiences with the monographic "Vicieae Database Project", Southampton and a floristic project, "La Flora de Veracruz", Xalapa, Mexico. I refer principally to the software used in these projects: CONFOR (Dallwitz, Chapter 23, this volume), EXIR (Brill and Estabrook, Chapter 5, this volume) and Pankhurst's identification program suite.

Since Shetler's (1974) attempt to "demythologize" taxonomic databases, significant advances have been made (Heywood, Chapter 1, this volume) but we cannot affort to ignore the dangers he referred to. Gibbs Russell and Gonsalves (Chapter 12, this volume) described the alienation of botanists whose first experience of databases was unsatisfactory. Shetler concluded that information retrieval in systematics was "still more potential than real" and that something was wrong with our designs and models or with our implementation of them. In this paper I look at taxonomic descriptions and at the contrasting uses to which we put them and ask if our models adequately reflect the complexity involved. I assume the desirability of taxonomic information systems to handle descriptive data but point to difficulties in implementing such systems.

THE NATURE OF TAXONOMIC DESCRIPTIONS

1. Features, Characters and Terminology

Conventionally, descriptions are prepared in a format which has evolved over centuries of taxonomic practice. Within this framework presentation may vary and, for each individual description, authors are free to choose whatever phraseology seems most appropriate. Special terminology has evolved simultaneously, enabling descriptions to be more concise, with space being a limiting factor in printed texts.

The quality of descriptions in the taxonomic literature has been criticized (Watson, 1971; Heywood, 1973). Pankhurst (1975) considered the effects of these shortcomings upon the construction of diagnostic data tables and concluded that new standards of taxonomic description would be needed. Computer-stored descriptions are created for a variety of tasks, including phenetic and cladistic analyses, and have more rigid formats than conventional descriptions; usually rectangular "taxon by character" tables. The use of more rigid formats has raised questions, previously ignored, related to the descriptive process itself.

I distinguish "features" of organisms from the "characters" employed by taxonomists to convey information about them (Allkin, 1979). Taxonomists examine individual organisms and conceive of features of interest (e.g. leaf length). But in order to record their observations and communicate with others they subsequently define characters: explicit statements to be evaluated by observation of specimens (e.g. the distance in mm between the petiole base and the point of attachment of the terminal pair of leaflets). For any one feature a host of different characters may be defined.

An often quoted advantage of "coded" (i.e. structured) descriptions is that they are "comparable". I am not always certain what this means but consider descriptions more valuable if:

(1) terminology is used consistently throughout the descriptions;
(2) the characters used to describe one organism are included in all other descriptions (homology permitting);

(3) each character consists of a set of mutually exclusive alternatives (or "states").

Fulfilment of the third and most exacting criterion is hampered by the use of specialist terminology. Arranging that states do not overlap is straightforward only for characters with single underlying features. This multistate character, for example, describes leaf length: "Leaf shorter than 7 cm; or 7 cm to less than 9 cm; or 9 cm or longer". Alternatively characters requiring more precise measurement may be defined quantitatively implying a large number of mutually exclusive states; the number of states being limited by the precision of observation possible.

Difficulties occur when characters are defined which describe more than one feature. The states of characters, representing feature complexes, cannot be logically ordered in a single dimension. I refer neither to quantitative ratios (e.g. length divided by width) nor to quantitative comparisons(e.g. twice as many stamens as petals), since these cases involve only one comparative feature which can be represented satisfactorily in a single dimension. Leaf shape, a common example of a feature complex includes the features leaf length, maximum leaf width and the relative position at which the leaf is widest. Leaf shape is frequently described by means of characters defined using terms such as "ovate" or "hastate". The states of such characters cannot be arranged in a logical sequence, nor are they necessarily logical alternatives. As already stated, specialist terminology provides a useful shorthand, terms having evolved to describe commonly occurring patterns of variation. The botanical terms applied to vestiture, for example, involve implicit observation of many features and individual terms deal with different subsets of them (Gómez-Pompa *et al.*, Chapter 14, this volume). The large number of features involved prevents any list of terms from including all possible variation patterns. Pankhurst (Chapter 25, this volume) describes these same problems occurring in the description of inflorescences and presents one alternative descriptive scheme in which the individual features are described separately. If comparable descriptions are desired, therefore, we must consider the possibility that, in the context of descriptive databases, much of our existing terminology may become redundant.

Computer-stored descriptions not fulfilling the third criterion do exist. Flora North America (FNA) employed a simple database design involving one list of descriptors and a second list of descriptor states (Pankhurst, 1983). The FNA Editorial Committee was "forced to choose between" a system designed to store "prefabricated descriptive phrases" for the purpose of forming keys or to store "elemental data describing the plants" (Shetler, 1975). The Committee selected the latter alternative "because it wanted to be able to deal in the traditional descriptive currency of the taxonomist".

It is a fallacy that structured character definitions employing logically alternative states are required only for identification (Allkin, 1980). To query a database one needs to know precisely what information has been stored: which features have been included and how each has been described. Without this information data searches become both lengthy and error prone. For example, to ask "Which beetles have red elytra?" I must know that the colour of beetle elytra has been recorded in the database, that the author of the database did not record as "vermillion", "rusty", or "scarlet" elytra that I would term "red", how partially red elytra have been recorded, etc.

Manual searches of descriptor lists are time consuming, but failure to consider all possibilities leads to errors. This can be avoided using predefined characters with discrete sets of mutually exclusive states, the definitions of which are made available to users of the database.

2. Descriptor Types and Comment Strings

Taxonomic descriptions involve a wide variety of descriptor types and one shortcoming of available data formats has been the restricted range catered for. In general, taxonomic data tables include both quantitative measurements (discrete and continuous) and multistate characters. Multistate characters may either be ordered or unordered, the latter being formulated from qualitative (presence/absence) features or from feature complexes. Ordered multistate characters describe quantitative features and may have quantitatively or subjectively assessed state boundaries (Allkin, 1979; examples are given in Table 1).

A further complication is that a taxonomist will wish to record personal opinions and, during data collection, observations of features not anticipated to be of interest at the outset of the study. He may also wish to include qualifying adverbs such as "remotely" or "deeply". Such information is best recorded as unformatted text, although this is not as reliably retrievable. If appropriate, these comment strings may subsequently be converted into a more accessible

Table 1. A classification with examples, of characters formulated from single quantitative features.

	Continuous Features	*Discrete Features*
Quantitative assessment	petiole length (mm.)	no of stamens
Quasi-quantitative assessment	petiole length: $\leqslant 5$ mm. or > 5 mm.	no. of stamens: $\leqslant 6$ or 7–9 or $\geqslant 10$
Subjective assessment	petiole: long or short	no. of stamens: few or many

form. Taxonomists readily manipulate information of both types: a database system should behave likewise. Happily, the DELTA data format (Dallwitz, 1980) provides this facility.

3. Variable Taxa

Taxonomists are not ultimately interested in observations of single specimens but are concerned with populations which, in general, show internal variation. Quantitative characters in particular vary and estimates of population parameters, such as the mode or range (Nauman, 1982), are necessary to describe a population adequately. Thus arises a need to record more than one value in a single cell of the rectangular data table.

Variable bistate characters can be relatively easily handled. Where only two states occur a third state "variable" can be used unambiguously. For a multistate character of three or more states "variable" becomes ambiguous and we must specify which combination of states occur. This is possible in the DELTA data format but not for retrieval programs such as EXIR (Adey *et al.*, Chapter 15, this volume), so that multistate characters must be manually broken down into a series of bistate characters. This can be cumbersome both during data capture and subsequent retrieval. Since in the EXIR "flat" structure, component bistate descriptors are not linked more intimately to one another than to other descriptors, information is effectively lost from the database.

Morse (1974) and Pankhurst (1976) use a coding scheme based on powers of 2, which partly overcomes these problems.

state no.	1	2	3	4	5	....	n
code value	1	2	4	8	16	....	2^{n-1}

Variable taxa are described using the sum of the appropriate coded values. Thus single values can unambiguously specify any combination of states. Code value 19 ($= 1 + 2 + 16$), for example, uniquely records the simultaneous occurrence of states 1, 2 and 5.

This technique is suitable for multi-state characters with a limited number of states. An upper limit is imposed on the number of states in some computers by the maximum size of an integer and, in any case, data collection becomes cumbersome and error prone for characters with more than a few states.

It is also possible to describe a variable taxon by recording more than one item description. Repeated descriptions are practical only if variation is limited since the number of additional descriptions becomes unmanageable where several characters vary simultaneously. Another desirable feature included in DELTA is that only variable characters need be described in the second and subsequent entries.

Additional complications occur where two or more variable characters show correlated distributions amongst individuals of a given taxon. Successful identification, particularly at higher levels in the taxonomic hierarchy, may depend on the recognition of such "complexed variation" and adequate description of the joint character distribution (Allkin and White, 1982). In Table 2 species A can only be distinguished from species B if the two subspecies of A are described separately.

Repeated taxon entry resembles taxonomic practice since new variants can be added during data capture as new specimens are observed. Unfortunately, the number of additional descriptions soon becomes unmanageable for highly variable taxa. The second

Table 2. An example of complexed variation.

	Leaf Shape	*Flower Colour*
Species A		
subspecies i	simple	blue
subspecies ii	compound	red
species	simple or compound	blue or red
Species B	compound	blue

alternative, explicit description of joint character distributions, is not possible in any existing data format but would sometimes be easier and more appropriate, allowing the taxonomist to incorporate observed correlation among characters directly.

At the cost of additional complexity in the data structure the latter alternative would enable us to query the database explicitly about such correlations. Coding effort would be reduced significantly since (i) more than one taxon might show similar joint distribution patterns, and (ii) the taxonomist could record either those combinations of character states which do occur or those that do not, whichever was the more convenient. Ideally, both mechanisms would be available. Data, entered in either format would be manipulated automatically from one to the other.

4. Probabilistic Information

Taxonomic data tables often include estimates of frequency and occurrence. Non-variable observations (e.g. "flowers, red") indicate that all individuals of the taxon always show that condition. For variable taxa (e.g. "flowers, yellow or white") the likelihood of each alternative state might also be recorded. Rarely is such information reliably available for higher organisms, although it is attainable more frequently for microorganisms and probabilistic identification routines exist (Payne, 1975). Subjective estimates of relative likelihoods are commonly included in publications concerning higher organisms by means of such terms as "usually", "rarely" or "frequently". Estimates of taxon abundance may also be available and, if in a quantifiable format, may be incorporated into identification routines. More normally, however, these are also recorded subjectively using phrases such as "rare" or "locally common". Unfortunately, such descriptors are rarely employed in a standardized fashion (Schmid, 1982) and may only be incorporated using comment strings attached to either individual observations or taxa. Such potentially useful information adds another dimension to the complexity of the data structure.

STRUCTURE WITHIN DESCRIPTIONS

Taxonomic descriptions are commonly presented as two-dimensional tables: taxonomists are accustomed to using paper, a two-dimensional medium, to record and present their data, and codified descriptions have been most often produced for phenetic analyses requiring observation of an entire set of characters for the given taxa. How effective a mechanism for data storage are such data tables? Many cells in such rectangular tables can be filled by intelligent interpretation of other cells. "Inapplicable", "variable" and "constant" data values provide no information that the taxonomist cannot obtain using other factual observations in the data table and his experience of taxonomic hierarchies and intercharacter dependencies. It is necessary to incorporate these logical relations within a database if efficient storage is to be achieved and if data retrieval is to prove as useful as intelligent visual searches of printed tables.

1. The Taxonomic Hierarchy

Conventional taxonomic classifications are stratified non-overlapping clusterings or hierarchies. Given descriptions of the objects at one level, the nested arrangement of classes allows us to make inferences about classes both higher and lower in the hierarchy: thus providing an important mechanism of data storage and retrieval. In the Vicieae Database Project for example, it would be desirable to print a description of the genus *Vicia* based upon the individual species descriptions contained in the database, or to remove the need to describe *V. galilaea* once complete descriptions of its two subspecies have been recorded. In monographic studies the relationships between taxa can be expressed in relatively few levels, but in floristic and faunistic projects such facilities assume far more importance. Existing data formats do not allow the taxonomic hierarchy to be included within the data structure and a powerful retrieval mechanism is thereby lost.

2. Serial Dependence among Characters

It is impossible to observe the length of bract hairs on plants with glabrous bracts: "bract hair length" is said to be dependent on "bract hair presence" and is inapplicable for those taxa without hairs. Among sets of characters several levels of dependence often occur. In Fig. 1, for example, "presence of bract hairs" is itself dependent on "bract presence", and that in turn is inapplicable for plants lacking leaves. Serial dependence takes the form of a directed graph, which for a given set of data may have branches with a variable number of arcs, a variable number of arcs leading from each node and a variable number of arcs converging on each node.

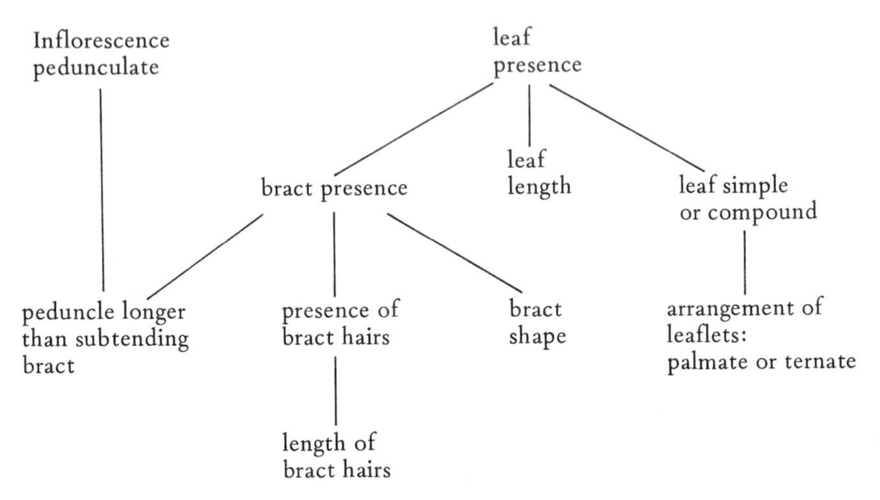

Fig. 1. An example of serial dependence among characters.

Explicit incorporation of serial dependence in a general-purpose data format could have many advantages other than those already manifest in key construction. For example:

(i) "intelligent" interactive data capture programs could prompt for appropriate observations as descriptions develop;
(ii) coding effort would be reduced, we would no longer need to score "inapplicable" observations;
(iii) automatic data verification would be enhanced;
(iv) considerable memory would be saved, provided that an appropriate internal data structure were employed;
(v) more readable printed descriptions could be produced; and
(vi) interactive identification and data retrieval would be more efficient.

During identification, the observation "bract hairs glandular" for example, could be interpreted to mean "bracts present", "bract hairs present" and "bract hairs glandular". Search algorithms could similarly be made more efficient. Without using serial dependencies, it would be impossible for a data-retrieval program either to detect logical errors in data queries or to respond to them intelligently.

In neither EXIR nor DELTA can such relationships be included, although such a facility is to be added to DELTA. Serial dependence, like the taxonomic hierarchy, assumes greater importance in floristic or faunistic studies than in monographic ones. Problems of homology are much severer when comparing a fern with an oak than when comparing two species within the same genus.

3. Data Models

Taxonomic descriptions do not have a "flat" (two-dimensional) structure: the taxonomic hierarchy, character dependence, taxon variability, missing data, estimates of frequency and personal opinions or additional notes all add further dimensions. Two approaches are possible: to "flatten" the data, forcing it into a two-dimensional mould (Brill and Estabrook, Chapter 5, this volume); or, since machine-stored descriptions need not necessarily be arranged in two dimensional arrays, to use the inherent structure of the data to provide a data-storage mechanism in which inapplicable data cells are avoided, characters consistent for a given genus need not be described for all species, and only those characters which vary need be described repeatedly for variable taxa.

The second alternative requires more sophisticated data management. One solution I am considering is the application of the "relational" database model (Barron, Chapter 3, this volume). I do not envisage a few large data files linked together but a system in which large numbers of small rectangular data tables are managed simultaneously. One table might contain details of character dependencies, another, observations of those characters applicable to all of the taxa at the highest level in the hierarchy. Yet another table might contain observations of a variable taxon for one, or possibly more, of those characters in which it varies. One way of evaluating these approaches is considered by White (Chapter 24, this volume).

THE MULTIPURPOSE DATABASE: CONFLICTING INTERESTS

In the proposed idealized taxonomic information system (Bisby, Chapter 2, this volume, Fig. 1) it is clear that the same body of information is intended for many distinct purposes. Indeed, one of the major advantages we claim for such systems is that descriptions need only be stored once, but may subsequently be employed for numerous tasks with intermediate data transformation being carried out automatically. Is this an over-simplification? In the remainder of my paper I examine some of the conflicting demands put upon descriptions intended for more than one purpose.

If descriptions are ultimately to be employed in identification, then a serious consideration during both character definition and data capture is whether or not descriptions should be strictly morphologically and developmentally correct. An author of a key must be at least as concerned with how the plant appears to a possibly inexperienced observer as he is with a pedantically correct description. For phylogenetic studies, however, the taxonomist will wish to record as ontogenetically correct a description as he is able. Such conflicting demands result in the need for multiple character definitions.

1. Sister Characters

Numerous characters may be defined from a single feature. These "sister" characters (Allkin and White, 1982) may differ in their scale of measurement, in the precision with which the character definition is worded, in the number of states and in the position of boundaries between adjacent states. As a result, two sister characters may have quite distinct properties and, in general, will be differentially suitable for a given role. For example, a reliably and readily observable bistate character, desirable for diagnostic purposes, may contain less information of classificatory value than a character with a greater number of less distinct states. Thus if data storage has several purposes, then it may be desirable to record more than one character per feature since no one character definition is necessarily the most appropriate in more than one role. Two multistate sister characters may be directly related: one being a contraction of the other with two or more states merged into one. In such cases only the most informative character need be stored in the database since the others can be derived from it when required. Sister characters need not, however, be so directly related and each may contain information not contained within the other. Then it is desirable first, to be able to use each character independently and, secondly, to ensure that sister characters be linked in such a way that only one, the most desirable, be used for any given application. Once again only one character need be stored in the database: the "master" character (Bisby and Nicholls, 1977) from which either sister character may subsequently be derived.

2. Character Definition

The desirability of characters using such subjectively assessed terms as "long" or "dense" is debatable. For identification, however, or whenever data is subsequently to be interpreted by anyone other than the author, such adjectives must be avoided. Their use requires an implicit comparison between the taxon described and others (presumably those under study) based on the observers experience. In addition to the problems of variable interpretation, characters using such terms are not suitable for use in identification since foreknowledge of the taxa concerned cannot be assumed of anyone needing a key.

A closely related database management problem is whether a database be built to be accessed directly by users or whether enquiries should be made via specialists manipulating the database. Character definition becomes far more complex if the database is made freely accessible to "end" users of the information service.

3. Multiple Character Sets

A desirable facility in any taxonomic database management system would be the possibility to include more than one character list. I now refer not to characters with distinct logical structure and a different information content but to characters differing only in the words used to define them or in the order in which words are used. The ability to replace one set of phrases with another has obvious application to the translation of descriptions or keys into other languages. The Vicieae Database Project has Portuguese and Spanish descriptor lists in preparation and Spanish descriptions of the Angiosperm families of Veracruz are printed using a modified version of Pankhurst's program DESC (Allkin, 1981a,b).

Less well appreciated is the need for different descriptor lists, written in the same language but phrased slightly differently, for different tasks. For example, if both printed descriptions and computer produced keys are to read adequately, then my experience has been that different versions of the character set are required. This is more easily resolved. Alternative character lists can be incorporated in the database, one for each application.

4. Character Set Formation

Distinct properties are sought of character sets intended for different purposes. For classificatory studies taxonomists seek highly correlated sets of characters: those characters whose states show similar distribution patterns among the taxa. For identification however, characters are selected for high discriminatory power and for unique information about the differences between taxa. The latter criterion conflicts directly with that expressed above as being desirable for classificatory character sets. Thus, character selection may proceed in quite opposite directions for classification and identification (Allkin, 1979, 1980). No single set of characters can be expected to be valuable for all roles. *A posteriori* character selection by the taxonomist (Bisby, 1970; Allkin, 1979) from the generalized character set included in the database, will be essential.

How frequently at present do taxonomists construct character sets with many purposes in view? Development of general purpose descriptive databases may require changes in the way taxonomists work but, since data would be made directly available to other scientists, it would reduce the time and effort wasted by taxonomists duplicating, rather than increasing, our inadequate knowledge (Heywood, 1978).

CONCLUDING REMARKS

Existing experimental projects manipulate coded taxonomic descriptions and already achieve many of their ends while exploring further potential advantages. Experience in these projects points to an exciting future but we are at an early stage. I have pointed to difficulties in the management of coded taxonomic descriptions which stem from shortcomings in traditional practice, the multiple and varied tasks for which we intend our data and failures in existing software to handle the complex structure of descriptions. To realize the full potential of descriptive databases, to provide more efficient handling of descriptions and to make the methodology available to taxonomists unfamiliar with computers, new software is needed. By discussing difficulties in existing systems I hope to provoke discussion. Our goals should be agreement on what is required of such software and progress toward a generalized data structure around which new systems could be built.

ACKNOWLEDGEMENTS

I thank Richard Pankhurst, Mike Dallwitz and all members of "La Flora de Veracruz" and "The Vicieae Database Project", particularly

Frank Bisby and Richard White, for discussion of the concepts presented here.

REFERENCES

Allkin, R. (1979). The evaluation and selection of plant characteristics for use in computer-assisted identification. PhD thesis, CNAA, London.

Allkin, R. (1980). "Some difficulties of computer-stored taxonomic descriptions". [INIREB document no. 80 30036.] INIREB, Xalapa, Mexico.

Allkin, R. (1981a). "Métodos para la identificación de las familias de Angiospermas de Veracruz. Informe No. 2. Descripción de la matríz de datos y su desarrollo". [INIREB document No. 81 30085.] INIREB, Xalapa, Mexico.

Allkin, R. (1981b). "Métodos par la identificación de las familias de Angiospermas de Veracruz. Informe No. 9. Las descripciones de las familias". [INIREB document No. 81 30089.] INIREB, Xalapa Mexico.

Allkin, R. and White, R. J. (1982). "Design criteria for a computer program to facilitate the acquisition, storage, retrieval and reformatting of biological descriptions". Southampton University Research Fund Report.

Bisby, F. A. (1970). Evaluation and selection of characters in angiosperm taxonomy. *New Phytol.* **69**, 1149–1160.

Bisby, F. A. (1984). Automated taxonomic information systems. *In* "Current Concepts in Plant Taxonomy" (V. H. Heywood and D. M. Moore, eds), pp. 301–322. Academic Press, London and Orlando.

Bisby, F. A. and Nicholls, K. W. (1977). Effects of varying character definitions on classifications of Genisteae (Leguminosae). *Bot. J. Linn. Soc.* **74**, 97–121.

Dallwitz, M. J. (1980). "User's guide to the DELTA system". [Division of Entomology Report No. 13.] C.S.I.R.O., Canberra, Australia.

Heywood, V. H. (1973). Ecological data in practical taxonomy. *In* "Taxonomy and Ecology" (V. H. Heywood, ed.), pp. 329–347. Academic Press, London and Orlando.

Heywood, V. H. (1978). Trends and priorities in plant taxonomy. *In* "Modern Trends in Plant Taxonomy", pp. 9–11. National Botanic Garden, Lucknow.

Morse, L. E. (1974). Computer programs for specimen identification, key construction and description printing. *Publ. Mich. St. Mus. Univ., biol.*, **5**, 1–128.

Nauman, C. E. (1982). Modal or mean character state values in plant descriptions? *Taxon* **31**, 716–717.

Pankhurst, R. J. (1975). Identification methods and the quality of taxonomic descriptions. *In* "Biological Identification with Computers" (R. J. Pankhurst, ed.), pp. 237–247. Academic Press, London and Orlando.

Pankhurst, R. J. (1976). "Standard taxonomic data format (Version 1)". Internal report, British Museum (Natural History), London.

Pankhurst, R. J. (1983). The construction of a floristic database. *Taxon* **32**, 193–202.

Payne, R. W. (1975). "GENKEY. Univ. Edinburgh Program Library Unit". Inter University/Research Councils Series, Report 24.

Schmid, R. (1982). Descriptors used to indicate abundance and frequency in ecology and systematics. *Taxon* **31**, 89–94.
Shetler, S. G. (1974). Demythologising biological data banking. *Taxon* **23**, 71–100.
Shetler, S. G. (1975). A generalized descriptive data bank as a basis for computer-assisted identification. *In* "Biological Identification with Computers" (R. J. Pankhurst, ed.), pp. 197–235. Academic Press, London and Orlando.
Watson, L. (1971). Basic taxonomic data, the need for organisation over presentation and accumulation. *Taxon* **20**, 131–136.

23 Automatic Typesetting of Computer-generated Keys and Descriptions

M. J. DALLWITZ

CSIRO Division of Entomology,
Canberra 2601, Australia

Abstract: Computer-generated keys and descriptions are usually produced on line printers, often in upper case only. To produce typeset versions, the computer output is edited and manually retyped at least once, and possibly twice. This paper discusses the problems of automating the process, and describes programs which have been used for the automatic production of a 200-page book on Australian grass genera.

INTRODUCTION

Programs for generating identification keys have been available for more than ten years (e.g. Hall, 1970; Pankhurst, 1970; Morse, 1971; Dallwitz, 1974; Payne, 1975). These programs read a set of taxonomic descriptions in the form of a matrix of numbers or codes, determine a suitable structure for a key, and print the key. Also, programs are usually available to convert the coded descriptions to natural language (i.e. an ordinary language such as English or French, as contrasted with the artificial, coded "language" in which the descriptions are presented to the computer). Although these programs have eliminated most of the labour of the actual construction of keys, the coding and correction of the data and the conversion of program output to the final, published form are still time consuming and tedious. The output, which is usually in upper case, would

Systematics Association Special Volume No. 26, "Databases in Systematics", edited by R. Allkin and F. A. Bisby, 1984, pp. 279–290. Academic Press, London and Orlando.

DATABASES IN SYSTEMATICS
ISBN 0 12 053040 6

normally be retyped before submission to the publisher, and then rekeyed for typesetting by the publisher. Proof-reading is required at each stage.

Over the last few years I have developed a data-coding system called DELTA, which has been designed for easy use by people rather than by computers. Programs for preparing keys and descriptions have been linked to a typesetting program, so that camera-ready copy can be produced without manual intervention. This paper describes some of the difficulties in the automatic production of typeset output.

THE DELTA SYSTEM

The DELTA system (DEscription Language for TAxonomy) has been fully described elsewhere (Dallwitz, 1980a,b). A brief description is given here to provide a background for the illustration of the typesetting process.

The taxa are described in terms of a set of characters. Figure 1 shows part of a set for describing grass genera (Watson and Dallwitz, 1981, 1982). The first part of each character is called the "feature". For multistate characters (e.g. character 2), the feature is followed by two or more "states". For numeric characters (e.g. character 4), the feature may optionally be followed by the units in which the values are recorded. The parts of the text enclosed in angle brackets are comments, which are printed in some contexts and omitted in others (see below).

Figure 2 shows a taxon description coded in terms of the character set of which Fig. 1 is a part. The first two lines contain the name of the taxon. This is followed by the "attributes" of the taxon. An attribute consists of a character number, a comma, and one or more state values. If there is more than one value, they may be separated by the symbols "/" (meaning "or"), "&" ("and"), or "–" ("to"). The symbols "V" ("variable"), "U" ("unknown"), or "–" ("not applicable") may be used in place of normal values. Attributes may also contain comments enclosed by angle brackets.

Some of the advantages of the DELTA format are:

(1) It is a "free" format, i.e. the data do not have to be placed in particular positions in the lines;

(2) there is no limit on the lengths of the character descriptions (in most formats, each feature and state is limited to one line of 70 or 80 symbols);

(3) the feature and state descriptions of each character are adjacent, and the states are numbered in a straightforward way;
(4) all the main types of character are catered for (unordered and ordered multi-state, integer and real numeric);
(5) the characters may be placed in any order (most formats require characters of the same type to be placed together);
(6) comments may be included in the character descriptions;
(7) there is no limit on the length of each taxon name;
(8) comments may be included with the taxon name;
(9) the name and the coded descriptions of each taxon are adjacent;
(10) distinction may be made between "known", "variable" and "inapplicable";

```
#1. <comments>/

#2. <longevity of plants>/
        1. annual or biennial <without remains of old sheaths
           or culms>/
        2. perennial <with remains of old sheaths or culms>/

#3. <whether culms woody or herbaceous>/
        1. culms woody and persistent/
        2. herbaceous <culms not woody, not
           long-persistent>/

#4. maximum diameter of culms/
        cm/

#5. culms <whether scandent or self supporting>/
        1. scandent/
        2. self supporting/

#6. <height of mature plants, whether 3m or more>/
        1. tall grasses to 3m or more high/
        2. grasses never reaching 3m in height/

#7. mature culms reaching <maximum height>/
        cm/

#9. <habit>/
        1. long-rhizomatous/
        2. long-stoloniferous/
        3. caespitose/
        4. straggling or decumbent, rooting at the nodes/

#10. <form of> rhizomes/
        1. pachymorph/
        2. leptomorph/
```

Fig. 1. Part of a character list in DELTA format.

(11) partial variability, i.e. the possession of more than one but not all of the states of a character, is coded in a straightforward way (some formats represent the presence of a state by the presence of the corresponding bit (binary digit) in a number, e.g. "state 1 or state 3" would be represented as "5");
(12) distinction may be made between "or", "and", and "to";
(13) comments may be included in attributes.

The program CONFOR (Dallwitz, 1980b) converts the coded data into natural language or into the formats required for various other

```
# [BNeyraudia] <Hook. f. 2 tropical Africa, Madagascar,
China, Indomalaya>/
2,2 3,1<cf. [IArundo]> 5,2 6,V 7,75–500 9<forming loose
tufts> 10,1 13,1 16,2 17,1 18,2 20,2 21,2 24,2 25,1<1.5[X–]4
cm, often longitudinally splitting> 26,2 27,2 28,2 29,2 30,1 32,1
34,3<long> 36,– 37,2 38,2 40,2 41,2 42,1 43,2 45,1 46,5<large>
47,1 48,– 49,– 50,2<more or less verticillate> 51,3 52,2 53,2
54,– 55,– 56,– 57,1 58,2 59,– 60,– 61,– 62,– 63,– 64,– 65,2 67,2
68,– 69,1 70,– 71,2 72,2<thin pedicels> 73,2 74,2 75,– 76,– 77,1
78,5–9 79,– 81,1 82,1 83,1 84,1<but glabrous except above the
abscission zone of each floret. The conspicuous hair tuft is on
the "callus"> 85,2<somewhat below the lemmas> 86,1<long
white hairs> 88,1 89,– 90,2 91,2 92,2 93,2 94,1<acute to
acuminate> 95,1 96,2 97,2 98,1 99,2 100,V 101,– 102,2 103,2
104,3 105,1–3 106,1–3 107,4–8 108,– 109,2 110,1 111,3 112,1
113,1 114,2 115,2<acuminate, larger than but similar to the
glumes> 116,3 117,1 118,2 119,2 120,2<membranous> 122,2
123,3 124,1<the distal florets with longer awns> 125,1 126,1–2
127,1/3<the lateral nerves may be excurrent into very short
awns> 129,– 130,1 131,– 132,1<curved> 133,1<bearded
marginally and on lateral nerves> 134,2 135,– 136,1<hyaline>
137,2 138,3 139,1 140,2 141,1 142,2 143,1 144,1 145,2 146,1
147,2 148,2 149,1 150,2 151,– 152,2 153,3 154,1 155,2 156,2
157,1 158,2 159,2 160,2 162,1 163,2 164,2 165,1<1.5[X–]2
mm> 167<grain angular> 170,– 188,2 189,– 190,1 191,1
192,2<the costals narrower and relatively longer> 193,1 194,1
195<and pitted> 196,2 197,– 198,– 199,2 200,2 201,2 202,2
203,2 204,2 205,2 206,2 207,1 208,2 209,2 210,2 211,1/2
212,1<mostly high domes> 213,2 214,1 215,1<and a few
solitary> 216,1<crescentic silica bodies> 217,1<but many of
the short-cells quite long> 218,1 220,2 221,2 222,1 224,2 225,2
226,1 227,1 228,2<bundle somewhat larger> 229,1 230,1 231,1
232,1 233,2 234,2 235,2 236,2 237,2 238,1<all bundles> 239,1
240,2<small groups of fibres occur abaxially under the
bulliforms> 242<2n = 40> 243,12
```

Fig. 2. A taxon description coded in DELTA format.

programs (for key construction, classification, etc). Figure 3 shows natural-language descriptions generated by CONFOR from the coded descriptions of Fig. 2. The figures in parentheses are character numbers, to permit easy reference back to the character list and the coded descriptions. The comments from the states in the character

Neyraudia <Hook. f. 2 tropical Africa, Madagascar, China, Indomalaya>.

(2) Perennial. (3) Culms woody and persistent <cf. *Arundo*>. (5) Culms self supporting. (6) <Height of mature plants, whether 3m or more> variable. (7) Mature culms reaching 75 to 500 cm. (9) <Forming loose tufts>. (10) Rhizomes pachymorph. (13) Culms branching vegetatively above. (16) Nodes glabrous. (17) Culm internodes solid, or spongy. (18) Fresh shoots not aromatic when crushed. (20) Plants unarmed. (21) Leaves not distinctly basally aggregated. (24) Leaf blades neither cordate nor sagittate. (25) Leaf blades broad <1.5–4 cm, often longitudinally splitting>. (26) Leaf blades not setaceous. (27) Leaves not needle-like, plants not prickly. (28) Leaves not pseudopetiolate. (29) Transverse veins very inconspicuous or invisible in the lamina. (30) Leaf laminae (or at least many of them) ultimately disarticulating entire from the sheaths. (32) Ligules consistently present. (34) Ligules consisting of a fringe of hairs <long>. (37) Leaf auricles absent. (38) Sheath margins free.

(40) Without pseudospikelets. (41) Plants bisexual with bisexual spikelets. (42) Having hermaphrodite florets. (43) Spikelets all alike in sexuality on the same plant.

(45) Inflorescence determinate (semelauctant). (46) Inflorescence paniculate <large>. (47) Overall form of panicle open. (50) Inflorescence neither digitate nor subdigitate <more or less verticillate>. (51) Spikelets not aggregated in heads, glomerules or spike-like units. (52) Inflorescence not leafy, not spatheate. (53) Inflorescence not comprising "partial inflorescences" and foliar organs. (57) Panicle branchlets capillary. (58) Not showing disarticulation of the spikelet bearing rachides. (65) Spikelets not all embedded.

Fig. 3. Part of the taxon description of Fig. 2, translated into natural language.

list (Fig. 1) are not printed in these descriptions (e.g. character 2), and the comments from the features are printed only if the attribute contains one of the special values "V", "U", or "–" (cf. characters 2 and 6). However, the comments associated with the attributes of a taxon (Fig. 2) are always printed (e.g. character 3).

TYPESETTING

For typesetting, I use an III COMp80 phototypesetter. This machine produces images on a high-resolution cathode-ray tube. These images are photographed on microfiche, 35 mm film, or A4-sized paper. A wide variety of type founts and special symbols are available, and the size and position of symbols are continuously variable.

The input to the phototypesetter is generated by a program, TYPSET (Dallwitz, 1980c). The input to TYPSET is a computer file containing, in addition to the text to be set, instructions indicating the required founts and layout. Instructions are enclosed in square brackets. The opening bracket is followed by symbols indicating what operation is required. For example, in the first line of Fig. 2, "[BNeyraudia]" indicates that "Neyraudia" is to be in bold type, and in the third line, "[IArundo]" indicates that "Arundo" is to be in italics. The results can be seen in Fig. 3.

In ordinary typesetting, all the typesetting instructions are entered manually. If program output is to be typeset automatically, many of the instructions must be inserted by the program producing the output.

AUTOMATIC TYPESETTING OF PROGRAM OUTPUT

1. Upper and Lower Case

A basic feature of typeset text is that it contains both upper-case and lower-case letters; therefore, the input material must contain indications of the required case. It is possible to use upper-case input, with the required case indicated by special symbols, but this places an unacceptable burden on the program user. Therefore, the input must be able to contain both cases.

(From the programmer's point of view, the handling of mixed-case text presents problems of program portability. Some Fortran compilers accept only upper-case letters. The solution adopted in CONFOR is to represent symbols by their ASCII or EBCDIC codes, and to use assembler input–output routines on those computers that do not support mixed-case input–output through READ and WRITE statements.)

If a character is at the beginning of a sentence in a natural-language description, it must normally start with a capital letter. If it is in the middle of a sentence (after "or" or "and"), it must normally start with a lower-case letter. If the first letter of the character in the character list is lower case, the program can change it to upper case if it is at the start of a sentence. The converse procedure, changing from upper case to lower case if the character is in the middle of a sentence, is not valid, because the upper-case letter may be necessary, e.g. in a proper noun or a chemical formula. Therefore, lower-case letters should normally be used at the start of each character in the character list, as in Fig. 1. There are further complications which must be taken into account by the program at the start of a sentence: brackets or typesetting instructions do not affect the case required for the subsequent letter; but other non-alphabetic characters (e.g. numbers, male or female symbols) require that the case of the subsequent letter should not be changed.

2. Typesetting Instructions

Figures 4, 5 and 6 show program-generated output containing typesetting instructions, and the corresponding output produced when this is processed by TYPSET. Figure 4 shows a DELTA-format description, Fig. 5 a natural-language description and Fig. 6 a key.

Some of the typesetting instructions in the program output are inserted by the program, but others are copied directly from the input files and must therefore be entered by the user when the input

```
[4H] [1.75`z][1~'z] [0i][0'i] ['"j]
['p]# [BNeyraudia] <Hook. f. 2 tropical Africa, Madagascar,
China, Indomalaya>/
[n]2,2 3,1<cf. [IArundo]> 5,2 6,V 7,75[X–]500 9<forming
loose tufts> 10,1 13,1 16,2 17,1 18,2 20,2 21,2 24,2
25,1<1.5[X–]4 cm, often longitudinally splitting> 26,2 27,2
28,2 29,2 30,1 32,1 34,3<long> 36,[X–] 37,2 38,2 40,2 41,2
42,1 43,2
```

\# **Neyraudia** <Hook. f. 2 tropical Africa, Madagascar,
China, Indomalaya>/
2,2 3,1<cf. *Arundo*> 5,2 6,V 7,75–500 9<forming loose
tufts> 10,1 13,1 16,2 17,1 18,2 20,2 21,2 24,2 25,1<1.5–4 cm,
often longitudinally splitting> 26,2 27,2 28,2 29,2 30,1 32,1
34,3<long> 36,– 37,2 38,2 40,2 41,2 42,1 43,2

Fig. 4. DELTA – format taxon description. Automatically generated input for the typesetting program (above) and output from the typesetting program (below).

data are being prepared. The instructions must be put in by the user if the program cannot know from the context that instructions are required. For example, the italic ([I...]) and bold ([B...]) instructions in Figs 4 and 5 are present in the input data (see Fig. 2). The instruction "[X–]", which appears several times in Fig. 4, produces an en-dash. This dash, which is longer than a hyphen (cf. character 3 in Fig. 1), is used to indicate a range of numerical values and the special value "inapplicable". The instructions in attributes 7 and 36 were inserted by the program, but the one in attribute 25 had to be included in the input data (see Fig. 2) because it is part of a comment.

[4H] [1.75`z][1~'z] [0i][3`'i] [j]
[1~'p] [BNeyraudia] <Hook. f. 2 tropical Africa,
Madagascar, China, Indomalaya>.
[p](2) Perennial. (3) Culms woody and persistent <cf.
[IArundo]>. (5) Culms self supporting. (6) <Height of
mature plants, whether 3m or more> variable. (7) Mature
culms reaching 75 to 500 cm. (9) <Forming loose tufts>.

Neyraudia <Hook. f. 2 tropical Africa, Madagascar, China, Indomalaya>.

(2) Perennial. (3) Culms woody and persistent <cf. *Arundo*>. (5) Culms self supporting. (6) <Height of mature plants, whether 3m or more> variable. (7) Mature culms reaching 75 to 500 cm. (9) <Forming loose tufts>.

Fig. 5. Natural language taxon description. Automatically generated input for the typesetting program (above) and output from the typesetting program (below).

The first line in each of Figs 4 to 6 were supplied by the program user and specify the overall features of the required layout, e.g. size of symbols, interline and interparagraph spacings, line and paragraph indentations, and tab positions. The second and third lines in Fig. 6 were also supplied by the user. CONFOR is able to place such lines anywhere except in the middle of individual character or taxon descriptions.

The rest of the typesetting instructions in Figs 4 to 6 were inserted by the program. "[n]" produces a new line, and "[p]" a new paragraph. "[1<]" invokes the first "left" tab, i.e. the left-hand end of the following text is placed at the tab position. "[1">...]"

invokes the first right tab, with a dot leader, i.e. the right-hand end of the text is placed at the tab position, and dots are placed between the previous text and the tabbed text (Fig. 6). If the tabbed text will not fit on the current line, a new line is automatically started.

```
[4H] [1.75`z][1.5~'z] [0'i][0`i] ['!][3.5`'!] ["!][160"!] [j]
[1~p] [c[BKey to the Grass Genera of New South Wales]]
[.5~p] [5`i]
[g][p][98g] 1(0). [1<]Rachilla prolonged beyond the
uppermost fertile floret. [1">2]
[n] [1<]Rachilla not prolonged. [1">133]
[g][p][98g] 2(1). [1<]Ligules membranous, not hair-fringed
(entire, jagged, tearing etc.). [1">3]
[n] [1<]Ligules of hairs, or hair-fringed. [1">76]
[g][p][98g] 3(2). [1<]Inflorescence a single, simple true spike.
[1">4]
[n] [1<]Inflorescence of narrow spike-like main branches.
[1">[BLeptochloa]]
[n] [1<]Inflorescence a spike of glomerules.
[1">[BElytrophorus]]
```

Key to the Grass Genera of New South Wales

1(0). Rachilla prolonged beyond the uppermost fertile floret.
. . . . 2
Rachilla not prolonged. 133

2(1). Ligules membranous, not hair-fringed (entire, jagged, tearing etc.). 3
Ligules of hairs, or hair-fringed. 76

3(2). Inflorescence a single, simple true spike. 4
Inflorescence of narrow spike-like main branches. . . .
. . . . **Leptochloa**
Inflorescence a spike of glomerules. . . . . **Elytrophorus**

Fig. 6. Automatically generated input for a typesetting program (above), and output from the typesetting program (below).

3. Line-Printer Output

Because typesetting facilities are at present relatively expensive and inaccessible, programs must still be able to produce acceptable line-printer output for day-to-day work. For most purposes, it is more convenient if the typesetting instructions from the input data are omitted from the line-printer output, because they make the output more difficult to read. CONFOR allows the user to specify whether these instructions should be left in or omitted.

4 *Requirements for Typesetting Programs*

TYPSET was written specifically for typesetting on the III COMp80 typesetter, but it is fairly easy to add options for output on other devices. Options for output on a line printer or a Sanders Media 12/7 typographic printer are already included.

CONFOR could easily be modified to insert typesetting instructions suitable for other typesetting programs (the instructions are inserted by a single subroutine). There are two important requirements for a suitable program. Firstly, it must have a tab instruction capable of automatically producing a new line if the tabbed text will not fit on the current line, as shown in Fig. 6. Secondly, its syntax must be fairly simple so that it is possible to write a subroutine to determine which symbols are typesetting instructions and which are not. This information allows CONFOR to omit typesetting instructions from line-printer output, and to skip typesetting instructions when finding the first letter of a sentence.

APPLICATIONS

The largest application to date of the automatic typesetting procedures has been the production of a book on Australian grass genera (Watson and Dallwitz, 1980). This book contains nearly 200 pages of descriptions and keys, which were produced and typeset automatically from a database. The cost of the computing and typesetting was about $ (Austr.) 800. Similar books for other regions could easily be produced from the same database. Descriptions of all the genera in the database are produced on microfiche at intervals of about a year and are available free of charge.

AVAILABILITY

My programs are available throughout Australia on the CSIRONET computing network, which is run by the Division of Computing Research (DCR), a Division of the Commonwealth Scientific and

Industrial Research Organization (CSIRO). Communication in the network is handled by about 100 minicomputers called "nodes", which are connected by lines leased from Telecom, the Government telecommunications organization. Nodes are located in DCR branches in the major cities, and in many other CSIRO Divisions, universities and non-CSIRO Government departments. Each node is equipped with a card reader and line printer, and many have other peripherals such as plotters. Terminals are connected to the nodes by private lines, leased Telecom lines, or dial-up lines, depending on the distance and frequency of use. The network has several "host" computers, which may be accessed from anywhere in the network, and a large number of "auxilliary" computers, which are run mainly as stand-alone computers and are connected to the network to allow the transfer of files to and from the host computers. The main host (a CDC Cyber 76) and expensive, specialized equipment such as the COMp80, are located at the main DCR site in Canberra. Output produced there is distributed quickly to other sites by courier services.

For use outside Australia, the programs are available on magnetic tape, free of charge. The standard tape format used is 9-track, 1600bpi, unlabelled, ASCII code, 80-character records, 12 records per block. Other formats can be supplied as necessary.

REFERENCES

Dallwitz, M. J. (1974). A flexible computer program for generating identification keys. *Syst. Zool.* **23**, 50–57.

Dallwitz, M. J. (1978). User's guide to KEY – a computer program for generating identification keys. *CSIRO Aust. Div. Entomol. Rept.* No. 4.

Dallwitz, M. J. (1980a). A general system for coding taxonomic descriptions. *Taxon* **29**, 41–46.

Dallwitz, M. J. (1980b). User's guide to the DELTA system – a general system for coding taxonomic descriptions. *CSIRO Aust. Div. Entomol. Rept.* No. 13.

Dallwitz, M. J. (1980c). User's guide to TYPSET – a computer typesetting program. *CSIRO Aust. Div. Entomol. Rept.* No. 18.

Hall, A. V. (1970). A computer-based system for forming identification keys. *Taxon* **19**, 12–18.

Morse, L. E. (1971). Specimen identification and key construction with time-sharing computers. *Taxon* **20**, 269–282.

Pankhurst, R. J. (1970). A computer program for generating diagnostic keys. *Computer J.* **13**, 145–151.

Payne, R. W. (1975). Genkey: a program for constructing diagnostic keys. *In* "Biological Identification with Computers" (R. J. Pankhurst, ed.), pp. 65–72. Academic Press, London and Orlando.

Watson, L. and Dallwitz, M. J. (1980). "Australian Grass Genera – Anatomy, Morphology, and Keys". The Australian National University, Research School of Biological Sciences, Canberra.
Watson, L. and Dallwitz, M. J. (1981). An automated data bank for grass genera. *Taxon* **30**, 424–429.
Watson, L. and Dallwitz, M. J. (1982). "Grass Genera: Descriptions". 4th edition (microfiche). Research School of Biological Sciences, Australian National University, Canberra.

24 | Implementing Small Database Systems with Specialized Features

R. J. WHITE

Biology Department, The University, Southampton SO9 5NH, UK

Abstract: A conventional database management system (DBMS) is a general-purpose program which manipulates data of predefined types, and converses with the user about his requirements in a specially constructed language. Some recent DBMS designs allow for the special needs of specific user groups by means of an additional program, which converses with the user and calls up the facilities provided by the data manipulation program. These types of DBMS are very large and complex program systems, although their capabilities are still limited by the underlying data structure. Instead of adding more layers of program between the data and the user to try to match the user's requirements with the predetermined data types and structure, the opposite approach is to put the user directly in control of his data. This can be done by providing one language which is used simultaneously to implement the DBMS, to act as the user interface and to contain the actual data. In this way the particular database and its DBMS become one entity. Such a system is extremely compact so that little memory is required on small computers, and is extensible so that it is easy to add, modify, or omit features according to the demands of a particular application. The paper considers this method of implementing specialized descriptive taxonomic databases using the language FORTH. The following design features are briefly considered: the hierarchical arrangement of both taxa and characters; dynamic redefinition of all taxon levels; characteristics of the command language; and fully interactive data entering and checking.

Systematics Association Special Volume No. 26, "Databases in Systematics", edited by R. Allkin and F. A. Bisby, 1984, pp. 291–308. Academic Press, London and Orlando.

DATABASES IN SYSTEMATICS
ISBN 0 12 053040 6

INTRODUCTION

1. Conventional Database Systems

(a) Single-level systems. The traditional approach to building a general-purpose database management system (DBMS) is to take an existing computer language and write a program which (i) manipulates data (structured into predefined data types), and (ii) converses with the user about his requirements. The latter interaction with the user is carried on in a specially designed language which is different from the language in which the system is written because the latter cannot conveniently express the users' requests. There is no resemblance between these two languages because each can only handle its own predefined set of data types.

Most of the complexity of the DBMS is caused by the need in (i) above to represent the data by means of the data types and structures handled by the implementation language, which may not be well chosen, and in (ii) to provide a complete interpreting system for handling the user language and coping with the errors and unforeseen circumstances which always arise when humans converse with computers.

A further source of difficulty and confusion for the user, especially one who is new to computers, arises because of the need for an operating system. In a conventional DBMS this is required for loading the program, providing access to files and causing it to begin execution. The operating system may also intrude in the dialogue between the program and the user.

(b) More complex systems. Some recent DBMS designs recognize that specific users or user groups have special requirements, so they add a second level of software. The DBMS proper is the first level, and provides a range of facilities to manipulate the data. These functions are not, however, designed to be used directly by the "end-user", i.e. the person who wishes to employ the DBMS to handle his data. Instead a second-level program is written in some language, either by the end-user or by a programmer seeking to provide for several similar applications. This language is often a well-known programming language such as FORTRAN or COBOL which has been "enhanced" for the purpose with extra features which give the programmer access to the capabilities of the DBMS. This program or

"interface" interprets the user's requests and calls up the facilities provided by the DBMS. However, the requirements of the user must still be constrained by what the DBMS can provide.

An outline of the structure of these types of database system is given in Fig. 1, and they are described in more detail by Barron (Chapter 3, this volume) and Freeston (Chapter 4, this volume).

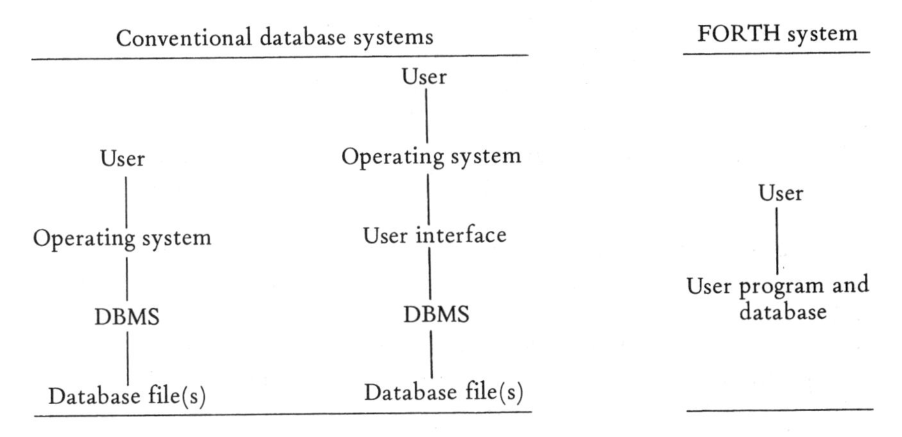

Fig. 1. Database architecture.

(c) Disadvantages of conventional systems. The disadvantages of the usual types of database system described above stem from the complexity of the programs involved. The amount of programming required to produce such a system represents a large investment, which can only be justified if a great deal of use is envisaged. A number of different types of use have to be anticipated in the design of the system, therefore such systems tend to be general-purpose.

Large systems involve additional overheads because of their very size and complexity. Further development of the system is difficult and the manuals may be lengthy and time consuming for new users to understand. Because of the number and diversity of users and the inexperience of some of them, the system must include comprehensive checking for errors in its use. The large size of the programs may necessitate special storage arrangements, such as overlaying, which add further to the complexity and slow down operation.

Unfortunately, a large general-purpose DBMS may tend to become even larger. Because the facilities provided are general purpose, they will not exactly suit the needs of many users. This leads to pressure for the provision of yet more facilities. TAXIR is growing in this way, as described by Brill and Estabrook (Chapter 5, this volume). An exception to this trend is provided by PRECIS II, which is significantly smaller than PRECIS I (Gibbs Russell and Gonsalves, Chapter 12, this volume), but this is a program for one database at a single institution and its design requirements were strictly delimited after experience with PRECIS I.

For biological systematists, there are two further difficulties. Systematists contemplating the use of database systems for descriptive purposes are not numerous compared with, say, the potential users of a business database system. Thus systematists do not generally form an effective lobby when a general-purpose DBMS is being designed or enhanced. TAXIR is an honourable exception to this rule.

The specialized nature of descriptive taxonomic data makes the previous problem more serious. The structure of a descriptive database has to take into account difficulties that do not exist for most other types of database. Allkin and White (1982) and Allkin (Chapter 22, this volume) have described these special properties of descriptive data. These characteristics are not all provided for in any general-purpose DBMS or in any specialized taxonomic database system.

2. Small Systems

(a) An alternative design approach. These disadvantages can all be eliminated or alleviated by reversing the usual design approach of increasing complexity. Small systems require fewer overheads in the form of purely organizational features and are easier to learn to use and cheaper to maintain. A small system could be aimed specifically at the needs of a particular user or group of users. For the taxonomic community, this may be the only way of implementing descriptive databases in the reasonably near future which take full account of the structure of the data.

Provided the fundamental design were sound, a small specialized system would be easy to modify as new requirements were identified. This modification could be carried out by users themselves, given reasonable experience or programming assistance. Thus the system would develop by "adaptive radiation" in the hands of many different users, user groups or institutions. Large conventional systems, on the other hand, are generally only created or enhanced, if they are enhanced at all, by professional programmers often working in software organizations remote from the users.

This paper is concerned with an attempt to outline the structure and advantages of such a small specialized system.

(b) Programming languages for databases. The aim in the present design is not to add more and more layers of program between the data and the user in an effort to match the user's requirements with the underlying data structure, but rather to put the user directly in control of the latter as in the right-hand system in Fig. 1. This can be done by considering the programming languages involved in providing a DBMS and making it accessible to users.

There are two main requirements: a language in which the database system functions are written, and a language to provide for the user to request those operations. Conventional statement-oriented languages, such as PASCAL and FORTRAN, are adequate though not usually ideal for programming the database functions. For the second requirement such programming languages are, however, entirely inadequate. This is because they are insufficiently interactive to permit the user to issue concise commands to cause a single action, such as counting the entries in a certain subset of the data, to be carried out immediately. The ability to give such commands is essential in interactive use of the database for testing, for exploration or validation of the data, or for on-line data retrieval.

The traditional ways in which such immediate response is obtained are either to design and implement an entirely new special-purpose interactive language in which the user gives his commands, or to write a single large program containing all the desired facilities, in which the user selects the operation currently required from a "menu" by typing a number or code name. Disadvantages of the former approach include the time spent writing a program to interpret the special language, the large amount of program storage required and the need for users who want to extend the system to learn two languages. In the latter case the obvious disadvantage is the complete inflexibility of the controlling program, no provision being made for individual users to develop their own set of operations for their special requirements.

In the present design the provision of a user interaction language is solved by writing the DBMS in a language which (i) can be extended or modified to handle the data types of interest; (ii) can be extended to contain the actual data itself, not merely the abstract data types involved; (iii) is suitable also as the user command language; and (iv) does not require the support of a separate operating system.

(c) Extensibility. No conventional computer language, not even those usually described as extensible, can satisfy all these requirements. Table 1 contrasts the types of extensibility available in such ordinary languages with the features of FORTH, a computer language of a highly unconventional kind which has recently attracted much attention. The extensibility of database systems written in the usual languages is also summarized. For reasons which will become clear later it is not necessary to distinguish between the language FORTH, and software, such as a database system, written with its aid.

Table 1. Extensibility of computer languages.

Type of Extension	*Conventional Languages*	*Database Systems*	*FORTH Databases*
New data objects or fields of predefined types	yes	most	yes
New combined data types	some	some	yes
New primitive data types	few	no	yes
New operators	few	no	yes
New language features	no	no	yes
Inclusion of the data itself	no	no	yes

Nearly all computer languages allow an indefinite number of new data items to be created according to the needs of the program. The analogous operation in a DBMS is the allocation of additional fields and records. In some languages and database systems these new objects must have one of a small number of predefined types, e.g. integral, real numeric or textual. In other languages and database systems, new types of data item can be formed by combining items of existing data types into "structures" or compound data objects.

Another way of introducing new types of data item that does not involve merely combining existing types into structures or records, is to establish completely new ways in which the pattern of bits of information stored in the computer memory will be interpreted. In most computer languages and database systems this can only be achieved by tricking the system into believing that the new types of data item are really just items of a standard type.

Other computer programming languages, including FORTH, are "untyped". This means that the system makes no assumptions as to the type of information to be stored in any memory location. The advantage lies in the complete flexibility allowed to the programmer in introducing new data types. The corresponding disadvantage is that an untyped language requires the programmer to take full control, with little automatic checking of the programmer's written code.

A few languages provide for new language operators to be written. A programmer might define a new data type for variable measurements as a pair of numeric data items, one for the minimum and one for the maximum value observed. To combine the ranges of different objects, it is natural to declare a new operator, which can be used as if it had been built into the language when it was designed. However, only in FORTH and related languages is this elegant property extended to all the features of the language. Thus a FORTH programmer can replace any facility of the language by an extension better suited to his needs.

The final type of extension considered here is that of extending the language to include the actual data. Thus the language itself, or rather one particular implementation of it, "knows" the information contained within the database. This permits a very natural and simple way to provide interactive enquiry and revision of the database. The integration of the DBMS and the data is also a very natural way to organize a system for a small computer where the DBMS does not have to provide for access to a number of completely different databases.

This amalgamation is only possible in a language which has such a high degree of extensibility as FORTH. Duplication of the DBMS in every database is unlikely to be a problem, even on microcomputers, because:

(i) the DBMS itself is very small (say 15K bytes) because of the extreme compactness of the FORTH code;
(ii) most users would only be involved in one or two databases;
(iii) on small computers, there would only be room for one or two databases anyway; and
(iv) each database need only contain those facilities it required.

THE PROGRAMMING LANGUAGE FORTH

The present proposal is not for a new form of data structure to challenge the existing rectangular, hierarchical, network or relational models, but a technique for implementing a database system. The software in the proposed system is based entirely on the use of the language FORTH, which has recently found a wide range of applications (Williams, 1980). A brief introduction to FORTH is given by James (1980) and more comprehensive guides by Brodie (1981) and Winfield (1983). Very small database applications written in FORTH, but not aimed at anything so complex as taxonomic data, have already appeared in Brodie (1981, pp. 328–340) and Filbey (1983). In order to explain how this language can help in the construction of a database system, it will be necessary first to present a short summary of its philosophy and features.

1. The Structure of FORTH

A FORTH language system can be considered as a collection of "words". Each word is actually a small independent piece of software. Typically there may be two or three hundred words each occupying ten to a hundred bytes of storage. A word has a name by which it is invoked, some action which it has been designed to perform and, perhaps, some storage space. Words may be named by the operator when working interactively and requiring the word to perform its action, or called within the definition of some other word.

Most words perform some useful action when they are invoked. These words are equivalent to the operators, procedure calls and other standard statement types of conventional languages. Each consists of a list of references to other words which constitutes a definition of the action of the word. Other words are equivalent to data items; they contain space for something such as text, a numerical value or a pointer to some other word to be stored and possibly altered.

Certain words are known as "defining words" and are responsible for the creation of new word definitions. These new words may be of any class including that of the defining words themselves: this generality contributes to the extreme degree of extensibility which is characteristic of FORTH.

FORTH provides simultaneously the following advantages, some of which have hitherto been considered incompatible:

(i) high execution speed;

(ii) very compact code with low memory requirements, thus leaving plenty of room for data storage and manipulation;

(iii) easy incorporation of short assembly language segments to allow, for example, efficient retrieval algorithms or use of appropriate display hardware;
(iv) structured programming, lacking a "goto" feature;
(v) reasonably easy portability of the software to other types of computer;
(vi) provision of an interactive user interface which makes the database easy to manipulate, even for inexperienced users;
(vii) the user can very easily combine, extend or replace the functions available to suit his own specialized application; and
(viii) there is no need to learn a separate operating system command language.

Sometimes an apparently awkward feature exists because it allows the user more freedom in the exploitation of other features. The lack of facilities such as arrays and strings which are widely held to be essential in a programming language is explained in this way. There is no point in providing a feature if it can be added easily by the user when he requires it. Thus a user who does not need the facility is not penalized by the extra storage requirement, nor is the feature provided automatically by the system in a form which may not be convenient for some applications.

2. FORTH in Use

Software written in FORTH consists mostly of the definitions of additional words. These are introduced into a FORTH program or "application" by means of defining words. For example, the defining word VARIABLE declares a new word which can be used to store a value in a fashion similar to that of other languages, as in

```
VARIABLE XYZ          (Declares XYZ)
```

This new variable can now be used in the following ways:

```
6 XYZ !               (Stores a value of 6 in XYZ)

XYZ @                 (Fetches the value of XYZ)
```

Note the use of comments introduced by the word "(". Arithmetic is performed on a stack using the reverse Polish notation familiar to Hewlett-Packard calculator users, as in

```
XYZ @ DUP *           (Fetches square of value of XYZ)
```

which fetches the current value of "XYZ", duplicates this value and then multiplies the two values together. The result is left on the stack for further manipulation.

A word can be defined to represent a sequence of commonly used words. This gives a facility analogous to a procedure in a conventional language. The new word is defined using ":" as follows:

: FRED existing words; (Defines a new word called FRED)

Words such as "FRED" can receive parameters to control their actions in several ways. Two common mechanisms are taking values previously placed on the stack, as used by the arithmetic operators, and reading the next string of text in the input, i.e. immediately following the word itself, as used by the defining word ":" to obtain the name of the new word. Defining words can themselves be created in various ways.

3. Extending the FORTH Kernel

The reason for stressing that the database system is FORTH-based lies in the way the language can be used for this type of application. FORTH can be used like any other programming language to write programs which do not themselves behave in the way FORTH behaves. However, a more economical approach is to extend the interactive system already present in FORTH. The "program" is then really just a collection of extensions to the original FORTH language system or "kernel". These extensions can be added in small stages, but at all times a working system is available.

A FORTH DATABASE SYSTEM

The remainder of this paper outlines some aspects of a small experimental database system which is being constructed to test whether the concept of a FORTH-based DBMS is practical. The system will provide a simple environment for the evaluation of alternative methods of implementing various features.

1. The Fundamental Parts of the System

(a) Conceptual access levels. The concept of levels of software, each level being a discrete part of the system which is translated by, or

makes calls to, lower-level software, is commonplace in conventional DBMS designs or other large software systems. FORTH programs are not strictly structured in this way because each word, when defined, is able to call upon any previously defined word. A word may also access a variable in which a pointer to a subsequently defined word is placed.

Nevertheless, it is convenient to distinguish loosely four broad classes of word in the present design, which may be thought of as constituting four levels of software. These are the lowest level of the FORTH language kernel itself, basic database manipulation words described in the following section, a third level which comprises the system command words and facilities as seen by the user and described in the distributed manual (should the system reach this stage) and, finally, the definitions added by a user for a particular database application.

(b) Database manipulation primitives. In FORTH jargon, a "primitive" is a word which carries out some elemental function. Which words are seen as primitives depends on the point of view of the programmer or user. In the present context, it is a convenient term to describe those second-level words which manipulate the data, but which are not employed directly by the user. They include:

(i) words for moving the current position within the database, such as selecting a given taxon, determining whether another taxon exists, moving to the next taxon or the previous taxon, determining whether a higher or lower level of a hierarchy exists, and moving to such a level;

(ii) words for restructuring the data, including creating a new taxon, removing an old taxon, or moving or copying a taxon, with corresponding words which operate on characters; and

(iii) words for manipulating the data, including words which fetch an existing value from the data lattice and which store a new value.

2. *Implementing Some Aspects of the Database System*

The properties necessary for an effective database system for taxonomists have been described at length by Allkin and White (1982) and only a brief summary is included here. They have been divided into two categories, those properties of the structure of the database itself and those more superficial, but none-the-less important, characteristics of the user command interface. Although the experimental system is not yet complete, it may be of interest to indicate the approach currently being taken in the implementation of some of these features.

(a) Data structure. A large number of features are desirable or essential if the internal data structure is to take proper account of the nature of descriptive taxonomic data. These are listed in Table 2 and some of them are discussed by Allkin (Chapter 22, this volume).

Table 2. Desirable data structure facilities.

Entering the data		
	1	from files and from keyboard
	2	data verification
	3	data addition and editing
Structure of the database		
	4	several character types
	5*	hierarchies of taxa
	6*	hierarchies of characters
	7*	additional dimensions, e.g. organs for chemical data
	8*	serial character dependence
	9*	variable observations accommodated
	10*	dynamic alteration of structure
	11*	associated information, e.g. comments and titles
Retrieval of information		
	12	by explicitly specified items
	13	by items which satisfy criteria
	14*	handling missing and variable data
	15	summary tables, reports and statistics
	16	conversion into various formats for subsequent analyses

*Not usually provided in conventional database systems.

Much flexibility can be provided in a database system if a general method is available for grouping both taxa and characters. Accordingly, such a method has a high priority in the consideration of the structure of the database. The present software is designed to allow for experimentation in this area.

A small group of FORTH words are responsible for handling the database structure. The details of the way in which the structure is handled are "known" only to these words. All other parts of the system with a need to refer to the structure of the database do so by calling them. When the structure is changed only these words need to be altered.

This class of structure-handling words falls within the group of "primitive" words described above. At present, they operate by linking the data cells in the approximate form of a two-dimensional lattice. Links exist pointing from a given cell to each of its four neighbours: above, below, to the left and to the right. Special arrangements deal with parts of the lattice which are missing for various reasons.

The taxa are linked into a tree which can, but need not, represent the accepted taxonomic hierarchy. A simple "tree-traversal" algorithm will convert the hierarchy into the linear order of the rows of the data lattice. The database access primitives are responsible for maintaining the current position, both in the hierarchy and in the data lattice. Alterations to the hierarchy can be made without changing the linear ordering of the data cells. Thus, for example, a species could be transferred to another genus and any subsequent retrieval commands would automatically reflect this new situation.

A similar arrangment is used to group the characters into classes for various purposes. It is hoped to provide for several taxon or character groupings to exist simultaneously in the same database. This is made possible because a given hierarchy does not require the data to be ordered in a corresponding way.

(b) Command language design. The design of the command language presents a much simpler problem than that of the internal database structure. However, this does not mean that little consideration need be given to the user commands. On the contrary, I believe the user interface is of vital importance to the success of the whole system, since it will largely determine the acceptability of the final product to the taxonomic profession. This is particularly so since many prospective users will have had little experience with computers. The desirable characteristics of computer languages are fairly well understood (Barron, 1977), and Table 3 lists them as they apply to database manipulation languages.

As in other areas, FORTH allows for experiment in the provision of user-command facilities. The command interpreter, which is already present in the FORTH kernel, has only to be extended by the addition of suitable words. Each word in the user interface, perhaps in association with other command words, is responsible for calling primitive words appropriate to the task in hand. The main design requirement is to harmonize two possibly conflicting considerations: first, to achieve a set of command structures which appears "natural" to a taxonomist; and secondly, to maintain simplicity by making the standard FORTH structures do as much of the work as possible.

Table 3. Desirable command language facilities.

Ease of use	
1	economy of concepts and syntax
2	uniformity and orthogonality
3	"glossary" and "help" commands
4	explicit error messages with echo of input and suggestion of possible remedy
5	editing erroneous commands without complete re-entry
Security	
6	protection of data from common errors
7	some facilities hidden from some users
Features for more experienced users	
8	repeat (i) the previous command
	(ii) use of previous item group
	(iii) use of previous term group
9	extensions: (i) user-defined commands
	(ii) command programs may be saved

For example, each taxon and character in the experimental database has a name which is the name of a FORTH word. A suitably constructed defining word is used to declare each new taxon name. Several different defining words might be used, one for each level in the taxonomic hierarchy, leading to a natural expression such as

SPECIES rosea

where "SPECIES" is such a defining word. It would read the following text to find the name of the new word about to be defined, in this case "rosea". However, one of the design requirements is that the taxonomic level of a taxon can be altered at will. Hence a construct such as

bridgesii BECOMES VARIETY

is likely to be needed. Note here that "bridgesii" is assumed to have been defined already and is being used as a word in the normal

FORTH way. The word "BECOMES" takes the last-named taxon and reallocates it to the taxonomic level specified in the following word. These two mechanisms could be combined as in

TAXON rosea BECOMES SPECIES

where "TAXON" declares a new taxon of any rank. New taxonomic ranks can, of course, be declared. Users concerned about the English language could define synonyms for these command words and hence write, in legal FORTH,

The_new_taxon rosea is_of specific rank

Further simple examples of the command language of the experimental database follow:

```
ITEM Rhipsalis TERM Areole             ( Selects a single entry )
  DIABLO PRINT                         ( and prints it          )

: LIST                                 ( Define a user's word   )
  TERMS[ Genus species ] PRINT;        ( to list binomials      )

SELECT Growth-form = jointed           ( All specified taxa     )
  AND Corolla-tube = absent LIST       ( are listed             )
```

3. Modifications to the System

Many advantages of using FORTH as the implementation language for a small database system should now be apparent. The most significant from the users' point of view will undoubtedly be the ease with which the system can be tailored for special requirements. At any stage words may be defined to perform new operations. Words added by the user or by other supporting staff fall into three classes according to the skill required to implement them.

Some words will simply be commonly used combinations of words selected from the user interface set, perhaps with additional standard FORTH words, and no special knowledge of the database structure will be required. These new words might be required, say, for adding a new type of output format. The next class of words consists of those which call directly upon the database manipulation primitive words. Some acquaintance with these primitives, not normally employed by the average user, will be necessary to write such definitions. They will be required only when a need for some completely new type of processing arises, such as the addition of a facility to sort the taxa into alphabetical order before printing. Finally, there will be those new words which implement a change to the internal structure of the database. These will typically be new versions of the database manipulation primitives, and will obviously require a full understanding of the workings of the present system.

Users can easily alter the amount of automatic checking that takes place. For example, on entering data, the system can be instructed to warn the user that a value has already been stored at the current location. This can be done by creating a word which performs the checking required and then providing the user with two words which switch the effect of the main word on and off. If the checking word is called indirectly via a vector or pointer it becomes easy for the user to substitute his own version of the word. This mechanism is required to stop previously-defined words from continuing to call the old version.

Other more advanced checking procedures that might be considered include provision to verify data, either as it is entered or when it is used, by the application of logical criteria specific to a particular database. For example, if a characteristic such as "leaflet number" has been defined, the likely range of the values associated with the descriptor could be specified. The system might then automatically query entries such as "−2", "4.6" or "green", or prompt the user for more information if a value was found to be missing.

These are really examples of a much more general characteristic. Within the one integrated system it is possible to provide several levels of interface: low-level facilities to be used only by experienced programmers; other facilities or commands designed for the average taxonomist who is aware of how the program is interpreting and carrying out his requests; a high-level "fool-proof" interface for users who are not aware of the consequences of certain commands; and perhaps a menu-driven interface which at every step gives the user a choice of two or three alternative actions.

4. *Practical Assessment*

An experimental version of the FORTH database system is being developed for two purposes: first, to test the feasibility of the FORTH database concept; and secondly, to enable different database structures to be evaluated and compared. An advantage of the proposed system is that the underlying data structure, or some part of it, can easily be changed without widespread alterations. This makes it very easy to experiment with alternative approaches to the data-storage structures and retrieval algorithms. It may not prove feasible to implement all the desirable features efficiently in one system. Even if no suitable data structure is found to enable their complete implementation, the system will be judged a success if it allows evaluation of such possibilities. These assessments certainly cannot be made easily by modification of existing conventional programs.

The ideas are being tested using a small experimental set of descriptive taxonomic data from some species, hybrids and cultivars in the subtribe Hylocereinae of the family Cactaceae. This set has been chosen to exploit the facilities provided, especially where the representation of the data in conventional database systems might cause difficulty.

In its present form the system is not suitable for a large practical application. It is not yet clear to what extent the present design could be extended to larger systems, although such a system represents a goal towards which work at Southampton is progressing. I hope to be able, in a further study with my colleagues, to report on a comparison of this new approach with more conventional software, such as that provided by relational database management systems. Such a comparison will involve at least two large-scale data sets drawn from practical descriptive taxonomic database projects: the Vicieae Database Project (Adey *et al.*, Chapter 15, this volume) and the *Flora de Veracruz* (Gómez-Pompa *et al.*, Chapter 14, this volume).

ACKNOWLEDGEMENTS

This work would not have been undertaken were it not for the stimulation provided by the Vicieae Database Project and by many discussions on the nature of taxonomic data and the problems associated with existing programs with: Margaret E. Adey, Robert Allkin, Mehmet T. Babac, Frank A. Bisby and Terry D. Macfarlane. Michael J. Dallwitz, George F. Estabrook and Charles R. Gunn have also helped me to formulate my ideas. None of those mentioned bears any responsibility for the heresies actually put forward.

REFERENCES

Allkin, R. and White, R. J. (1982). "Design criteria for a computer program to facilitate the acquisition, storage, retrieval and reformatting of biological descriptions". Southampton University Research Fund Papers, Southampton.

Barron, D. W. (1977). "An Introduction to the Study of Programming Languages". Cambridge University Press, Cambridge.

Brodie, L. (1981). "Starting Forth". Prentice-Hall, New Jersey.

Filbey, G. (1983). FIG database. *Forthwrite* 2 (5), 13–14. Forth Interest Group UK.

James, J. S. (1980). What is Forth? A tutorial introduction. *Byte* 5 (8), 100–126.

Williams, G. (1980). Threads of a Forth tapestry. *Byte* 5 (8), 6–10, 128–134.

Winfield, A. J. (1983). "The Complete Forth". Sigma Technical Press, Wilmslow.

25 | On the Description of Inflorescences

R. J. PANKHURST

Botany Department, British Museum (Natural History), London SW7 5BD, UK

Abstract: During the construction of a pilot floristic database it was necessary to choose a set of consistent and comparable characters in order to describe a wide diversity of different inflorescences. This can be done by using the characters which underlie the conventional classification of inflorescences. Examples are given where the scheme is applied to unclassifiable types of inflorescence.

INTRODUCTION

During a pilot project to examine the feasibility of constructing a floristic database (Pankhurst, 1983) it became apparent that the description of inflorescences presented particular problems. Other groups of characters, such as those describing trichomes (hairs) and the fruit, also cause similar, but not so severe, problems. This paper proposes a rationale for describing inflorescences, and is a companion to Pankhurst (1983).

The database included a relatively small number of species selected for as wide a morphological diversity as possible, i.e. one species for each of 50 orders. Because the intention was to allow the stored data to be useful for the construction of identification keys to distinguish families and for the comparison of taxa in detail, descriptions needed to be comparable, consistent and complete. It would be possible to store something roughly equivalent to the text of a conventional Flora in a database for subsequent retrieval of data, but this would

Systematics Association Special Volume No. 26, "Databases in Systematics", edited by R. Allkin and F. A. Bisby, 1984, pp. 309–320. Academic Press, London and Orlando.

DATABASES IN SYSTEMATICS
ISBN 0 12 053040 6

perpetuate serious descriptive ambiguities and entail the loss of potentially useful information. In addition, the characters used need to have exclusive states and the logical dependencies of characters upon one another have to be made explicit. These are all desirable properties of taxonomic information whether or not computers are used. The logical dependencies were expressed as conditions that showed which states the controlling characters needed to have to enable the controlled characters to occur. Characters were defined each with a list of appropriate states which could go with them, and this list of characters was effectively a standard set to be used for all taxa. The data format used was the same as that already in use for a set of programs for key construction and allied problems (Pankhurst, 1978). Because not all the taxa in the database show anything like the complete range of standard characters, a questionnaire program was written so that a user could enter descriptions of new taxa in an efficient way. It is rather impractical to achieve a high standard of accuracy, consistency and comparability in floristic and monographic publications without the aid of computers. Watson (1971) gives a criticism of the quality of existing taxonomic information in literature.

Conventional descriptions of inflorescences involve something more than the detailing of characters and their states. When it is stated that a "raceme" is present, use is being made of a prior classification of types of plant structure. This is, of course, also true when fruit, trichomes and other organs are being described. The standard classification of inflorescences has long been known to have unsatisfactory features. Bentham (1858, p. 11), after defining types of inflorescence, goes on to say:

> There are numerous cases where inflorescences are intermediate between some two of the above, and are called by different botanists by one or the other name, according as they are guided by apparent or theoretical similarity.

Examples of this are not difficult to find. The inflorescence of the *Rubus fruticosus* aggregate has either simple lateral branches on a central rachis, or compound lower branches. The central axis has a terminal flower which flowers first. British authors (Rogers, 1900; Watson, 1971) call the compound inflorescence a "panicle", which is incorrect since a panicle does not have a terminal flower, although it has the same branching pattern. They describe the simpler inflorescence as "racemose", by reference to the branching, which is reasonable but not a complete description. Clapham *et al.* (1962) avoid giving the *Rubus* inflorescence a name except to call it "cymose" for those forms which are flat-topped, which is again correct but not a complete description.

No attempt has been made to create a new classification of inflorescences, nor to introduce any new terminology. There have been many attempts to improve the classification and many new terms have been coined, e.g. Troll (1964) and Weberling (1981), but it is not necessary to do either of these to create a workable descriptive scheme. All that is needed is to record the characters which underlie the familiar classification. Also, no attempt has been made to describe inflorescences through knowledge or theories of how they develop, nor to express any ideas about how they evolved, but simply to record their actual morphology. At the outset it seemed that there would be an almost infinite variety of ways in which a consistent and comprehensive scheme for the description of inflorescences could be designed, but in practice it proved quite difficult to find even one workable scheme. The scheme described below is not intended as a new standard, but rather as an exploration of the difficulties and some of their solutions. It is definitely not complete, since only one-third of world angiosperm families were considered. Any characters which happened to be constant in the sample chosen were not included, merely in order to reduce the number of characters. Definitions of the classical inflorescence types were taken from a good modern reference as a standard (Lawrence, 1951). Other useful sources were Gray (1879) and Radford *et al.* (1974).

DEFINITION

The term "inflorescence" has been used in two different senses (i) as a cluster of flowers, and (ii) as the structure in which the flowers are arranged. Some kind of compromise between these two senses is needed, so as to be able to say that all flowering plants have inflorescences. There is a problem over whether or not an inflorescence can be said to be clearly differentiated from the supporting stem structure. One might expect to say that a differentiated inflorescence has some obvious character which is not shared by the stem structure beneath. It is not enough to say that in the inflorescence the flowers are grouped since, e.g. in *Malva sylvestris*, they occur in drooping clusters and yet the inflorescence is normally considered to be the entire flowering structure. One might also expect that a differentiated inflorescence has a different geometrical structure from that of the stem, but there are cases where both have the same, e.g. *Viscum album*. While no objective definition of this has been found for the moment, the distinction seems to be easily applied in practice. There is, therefore, a character which distinguishes inflorescence in sense (i) above from sense, (ii) i.e. not differentiated and differentiated, respectively.

If the inflorescence is undifferentiated, then it is defined as follows:

(a) if the plant has only one flower, then that flower is the inflorescence (the trivial case);
(b) if the flowers occur singly, but there is more than one, then the inflorescence begins at the lowest common branch point;
(c) if the flowers occur in clusters, but all the clusters are equivalent in structure, then each cluster is an inflorescence, beginning at the base of the cluster; and
(d) if the flowers are in clusters, but the clusters are not all equivalent, the inflorescence begins at the lowest branch point which is common to all clusters.

CHARACTER SET

The first character to record is whether the inflorescence is differentiated, and the next is obviously the type of inflorescence, (a) to (d); unless the inflorescence is of the differentiated kind. The definition corresponds to the conventional notions of what an inflorescence is, but with some exceptions, e.g. in the case of a characteristic taxon of the Asteraceae with a capitulum, each capitulum will be an inflorescence. Other characters may then be needed to describe the arrangement of the inflorescences as a whole.

An obvious property of an inflorescence is the degree to which it is branched. It is natural to call any side branch a first-order or primary branch and so the main axis, if there is one, could be regarded as the branch at order zero. It is clear that inflorescences differ in their degree of branching, and also in the extent to which branching continues. A raceme (Fig. 1a) has just one order of branching, whereas a panicle (Fig. 1b) branches more or less without limit. Characters are needed to describe whether the degree of branching is limited or unlimited, and, if it is limited, the number of orders of branching which occur. The existence of unlimited branching makes it convenient to number the orders of branching from the central axis (order zero) and so on to higher-order branches, rather than starting counting from the ultimate flowering branch up towards the main axis. It is possible to have limited branching to more than one order, as in *Geranium molle*, Fig. 1c. Here the lower branches are of order 2 and the upper of order 1. With either kind of branching, flowers do not necessarily occur at each order, e.g. in *Euphorbia peplus* flowers only occur at order 3 and beyond. In fact, this species

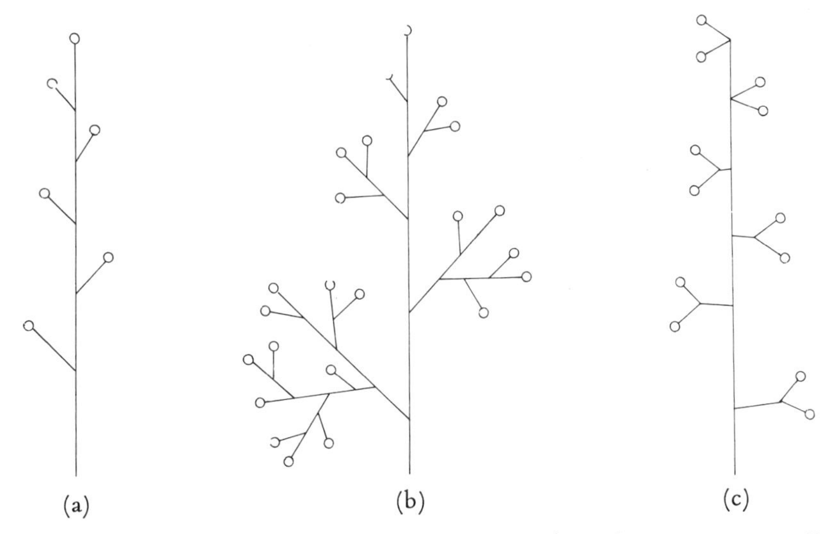

Fig. 1. Inflorescence diagrams: (a) raceme, (b) panicle, and (c) *Geranium molle*; individual flowers are shown as open circles and leaves (bracts) are omitted.

has a special structure called a cyathium in which rudimentary flowers of different sexes are associated in an involucre. For present purposes, this has been ignored except for counting the extra branching involved. Hence, the order of branching at which flowers first appear is another character to note. The form of the branching, e.g. dichotomous, may not be the same at each order, and so there is a character for form of branching at each order. An example is *Euphorbia peplus* again, where the main stem divides in an umbel at the apex, and may have lateral alternate branches as well, but branches at higher orders are dichasia.

The traditional notion of determinate or perfect inflorescences as opposed to indeterminate or imperfect ones causes some difficulty. Partly this character expresses whether or not the inflorescence has a terminal flower, which is perfectly straightforward. However, it is possible for an inflorescence to lack a terminal flower on the main axis, but to have one on higher-order branches (Weberling, 1965). Hence, the presence or absence of a terminal flower needs recording at each order of branching. The other part of its meaning concerns the order of flowering. If an inflorescence is limited with one order of branching, then the order of flowering, from the top down (or centrifugal, or centre outwards) or from the bottom upwards (or centripetal, from the periphery inwards) will be easy to observe. If the branching goes on without limit, then one may expect the branches which grow first to flower first, but the order of flowering may not be obvious unless only branches of the same order are compared. There is also the possibility that flowers which are all at the same order flower simultaneously, e.g. *Euphorbia peplus*, or apparently at random, e.g. *Dipsacus*. In order to be quite certain of the flowering order, observations of living plants over a period are necessary. These have not been carried out, so characters of flowering order have been omitted for the present.

Another character used to describe inflorescences is whether or not the flowers are stalked. This is one of the differences between a spike and a raceme. In the present database, all the taxa included have inflorescences with flowers which are either all sessile or all stalked, and so this character is included to describe the complete inflorescence. However, inflorescences do exist which have both sessile and pedicellate flowers together e.g. in the Loranthaceae (Kuijt, 1981) and a more comprehensive database would also have to be able to describe these.

The notion of an umbel implies not only a group of flowers whose pedicels all come from a common point (a fascicle) but also that these flowers are held erect and, possibly, that all the flowers are at the same level. Hence, at least three characters are expressed at once here. In order to describe other fascicles in a comparable fashion, umbels are treated as a kind of fascicle.

The terms "monopodial" and "sympodial" are used here to describe two principal modes in which inflorescences can branch, namely, with a principal axis or without, respectively. Nothing more is implied by the use of these terms other than geometrical description. Other kinds of branching are included within them, e.g. monopodial branching types include opposite, whorled, etc., and sympodial types include dichotomies, dichasia, etc. The helicoid cyme is an interesting case since it has been shown to be derived from a dichasial form of branching with the branches missing on one side. However, in keeping with the guidelines given above, and going on appearances only, it is treated as a monopodial form since it does have a main axis.

It is common practice when describing inflorescences to state that the flowers are axillary, or terminal, or both. If the presence or absence of terminal flowers, the occurrence of flowers at stated orders of branching, and whether flowers are stalked or not are recorded, then the same information is already covered.

Table 1 shows the classification broken down into a set of component characters. This analysis is not the only possible one, and refers to a particular set of definitions (Lawrence, 1951). The definitions of "raceme" and "panicle" assume that the flowers occur singly. This is not explicit in the written definitions, but the illustrations show single flowers only, as does Gray (1879), and this appears to be normal usage. Table 1 also lists the fundamental set of characters which were used to describe inflorescences.

There are certain characters which have deliberately not been included. For example, there is no character called "inflorescence type" with states such as "raceme, panicle, cyme", etc. This is because these familiar terms are neither necessarily exclusive of one another, consistently defined, nor completely comprehensive. These shortcomings have long been recognized, e.g. Rickett (1955). Efforts to resolve them have usually led to attempts to redefine or extend the classification of inflorescences. What has been done here is simply to find the characters which underlie the classification and to record those, so that the classification has been neither abandoned nor revised. Although terms such as "raceme" were not stored explicitly in the database, they can easily be deduced from the data about the component characters. There would be no difficulty in adding characters such as "type" = "raceme" or "not a raceme" as required, except that inflorescences occur quite frequently which will not fit into any of these categories, as in the examples given below. Once the actual morphology has been rigorously described, the data can then be used for studies of floristics, evolution or development, or for whatever purpose is desired.

Table 1. Analysis of characters of inflorescences.

	General							1st-order branches					
	Limited or unlimited	No. of highest definite order	Order no. of first flowers if indefinite	Shape	Flowers stalked	Main axis terminal flower	Axis straight or curved	Branches present	Flowers present	Clusters present	Branching type	Monopodial branching type	Sympodial branching type
RACEME	L	1	X	E	+	—	St	+	+	—	M		X
PANICLE	U	X	1	E	+	—	St	+			M		X
CORYMB				FT		—							
CYME				FT		+							
UMBEL				FT	+	—		+		+	S	X	F
SPIKE or SPADIX	L	1	X	Cy	—	—	St	+	+	—	M		X
CONCINNUS + HELICOID CYME						+	C	+	+		M	Os	X
SCORPIOID CYME						+	C	+	+		M	A	X
SCAPE	L	0	X		+	+		—					
DICHASIUM	U	X	1			+		+	+	—	S	X	Di
CAPITULUM	L	1	X		—	—			+	—	S	X	H

Key: + present, — absent, X inapplicable, A alternate, C curved, Cy cylindrical, Di dichasial, E elongated, F fascicle, FT flat-topped, H head, L limited, M monopodial, Os one-sided, S sympodial, St straight, U unlimited.

BRACTS

A bract is a leaf which occurs in the inflorescence, often reduced in size by comparison with other leaves, and often occurring in association with flowers, according to Lawrence (1951). There are in fact two quite different types of bract which need to be distinguished and described separately:

(1) bracts which are generally distributed throughout the inflorescence;

(2) bracts which occur in association with a flower or flowers.

These might occur singly on a pedicel, in groups (involucre or epicalyx), or in more than one place.

It is also possible for both kinds of bract to occur together, as in *Malva sylvestris*, where leaf-like bracts subtend the branches of the inflorescence and where individual flowers also have an epicalyx. Further, if the flowers are unisexual, the male and female bracts may be different, e.g. *Fagus sylvatica*, and require separate description. A set of characters to describe this situation is shown in Table 2.

A bracteole or bractlet is a secondary bract. There is no accepted convention as to whether this means a bract on a greater order of branching, the second kind as opposed to the first, or to the second row of bracts on a pedicel. All these different meanings can be found in the literature. For the sake of clarity, in the database a bracteole is taken to mean a bract at a further order of branching. Bracts do not

Table 2. Bract description scheme, list of characters

Characters
Bracts present/absent
Bracts (type 1) like stem leaves/or not
Bracteoles present/absent
Bracteoles like bracts/or not
Bracts type: generally distributed (type 1)/with flowers (type 2)/or both
Bracts (type 1) occurring from branching order, no.
Bracts (type 1) no.
Bracts (type 2) at branching order, no.
Bracts (type 2), no.
Bracteoles associated with flowers/or not
Bracteoles at inflorescence branching order, no.
Bracteoles, no.
Bracts (type 2), length relative to calyx
Epicalyx present/absent

necessarily occur at all orders of branching, or on successive orders, so characters are needed to say at which order bracts and bracteoles occur (see Table 2).

EXAMPLES

Two examples of descriptions are shown in Fig. 2. These have been transformed to text form from the data in the database. Brackets indicate information not actually stored in the database. This was only because the database is experimental and rarely used characters have been left out for the time being.

Geranium molle

Inflorescence monomorphic, terminal, flowers in differing groups, stalked, hermaphrodite. Inflorescence cylindrical in outline, branching to limited orders 1 or 2, without a terminal flower. Main axis straight, monopodial, branching alternate. 1st order (upper) branches present, with flowers in fascicles (pairs). 1st order branching (lower) sympodial, dichotomous, 2nd order branches present with flowers in fascicles (pairs).

Bracts present, generally distributed, occurring singly from order 0, unlike stem leaves. Bracteoles present, unlike bracts, not associated with flowers, occurring in pairs from order 1.

Euphorbia peplus

Inflorescence monomorphic, terminal, flowers in equivalent groups, stalked, regularly monoecious. Inflorescence obconical in outline, branching without limit, flowers first appearing at order 3, without a terminal flower. Main axis straight, branching sympodial and fascicled (umbel) or (sometimes) monopodial and alternate (as well). Ist order branches present, without flowers. 2nd order branches present, without flowers, branching sympodial and dichasial. (3rd and higher branches similar, with flowers).

Fig. 2. Examples of inflorescence descriptions.

1. Geranium molle

Much of the data for the database was compiled from a standard Flora (Clapham *et al.* 1962). For this species, it simply states that flowers occur in pairs, and nothing more. In fact, the inflorescence is unclassifiable in conventional terms but has an uncomplicated structure, Fig. 1c. It has no terminal flower, and the branching is limited, to second order below and first order above. It cannot be called a raceme because it branches beyond first order. It cannot be called a panicle either, because the branching is limited. Also the flowers are in pairs rather than solitary, which does not fit either definition. The description of the bracts, however, is straightforward.

2. Euphorbia peplus

In this case the Flora states "inflorescence cymose, the primary branches umbellate". This is more informative than the last example, but misses the point that the cymes are dichasial and that the actual flowers do not appear until the third order of branching. This is once again an unclassifiable inflorescence. The description of the bracts is not included in Fig. 2 since it is very similar to that of the above species.

It is apparent that the inflorescence descriptions given above are both more precise and more lengthy. They are intended to be stored in a computer and recalled or compared as required, rather than to be printed and published. Recording these descriptions is not laborious since a computer program has been written to help (Pankhurst, 1983).

CONCLUSIONS

It has been shown that it is possible to construct a descriptive character scheme for inflorescences which will be comparable and consistent, and with exclusive character states. This scheme is suitable for creating computer databases for floristics and can cope with unclassifiable types of inflorescence. This is achieved by identifying the characters which underlie the conventional classification, so that the classification does not need to be revised or extended.

The character scheme will need to be tried out on a wider range of taxa and extra characters introduced to cope with the extra diversity. The expression of the characters already included could probably be improved. There must be other satisfactory ways of setting up character sets for describing inflorescences, and these should be explored and evaluated.

ACKNOWLEDGEMENTS

Dr N. K. B. Robson gave much helpful advice on the meaning of family characters, and Leslie Watson provided much valuable criticism.

REFERENCES

Bentham, G. (1858). "Handbook of the British Flora". Lovell Reeve, London.

Clapham, A. R., Tutin, T. G. and Warburg, E. F. (1962). "Flora of the British Isles". 2nd edition. Cambridge University Press, Cambridge.

Gray, A. (1879). "The botanical textbook. Part I. Structural Botany". Ivison, Blakeman, Taylor and Co., New York and Chicago.

Kuijt, J. (1981). Inflorescence morphology of Loranthaceae – an evolutionary synthesis. *Blumea* 27, 1–73.

Lawrence, G. H. M. (1951). "Taxonomy of vascular plants". Macmillan, New York.

Pankhurst, R. J. (1978). "Taxonomic data format, version 2". Internal document, Botany Department, British Museum (Natural History), London.

Pankhurst, R. J. (1983). The construction of a floristic database. *Taxon* 32, 193–202.

Radford, A. E., Dickison, W. C., Massey, J. R. and Bell, C. R. (1974). "Vascular plant systematics". Harper and Row, New York.

Rickett, H. W. (1955). Materials for a dictionary of botanical terms, III. Inflorescences. *Bull. Torrey Bot. Club* 82, 419–445.

Rogers, W. M. (1900). "Handbook of British Rubi". Duckworth, London.

Troll, W. (1964). "Die Infloreszenzen. Typologie und Stellung im Aufbau des Vegetationskorpers. 1. I Deskriptive Morphologie der Infloreszenzen." Gustav Fischer, Jena.

Watson, L. (1971). Basic taxonomic data: the need for organization over presentation and accumulation. *Taxon* 20, 131–136.

Weberling, F. (1965). Typology of inflorescences. *J. Linn. Soc. (Bot.)* 59, 215–221.

Weberling, F. (1981). "Morphologie der Blüten und der Blütenstände". Ulmer, Stuttgart.

26 Databases in Systematics: A Summing Up

G. Ll. LUCAS

Royal Botanic Gardens, Kew, Surrey TW9 3AB, UK

Abstract: I draw attention to three advances since the "Computers in Botanical Collections" conference of 1973: the coming of pragmatism, the appearance of microcomputers and the availability of user-friendly software. In addition there are two of the recurring themes at this conference which I wish to emphasize in their importance to systematists. The first is that text-processing facilities reduce the time and cost of publication. The second is that computerized databases, such as that projected for *Index Kewensis*, will provide an undreamt of variety of indexing facilities. I believe we now have the tools to resolve what ten years ago were impossible dreams.

SUMMING UP

Three main advances are apparent to me since the 1973 Kew Conference which led to the publication of *Computers in Botanical Collections* (Brenan *et al.*, 1975). The first reflects the very healthy move away from the grandiose hypothetical, with the overriding emphasis on the need to justify why we needed computer aids and "what we could do with them in theory". This conference reflects a range of excellent pragmatic approaches actually resolving problems, candidly exposing both our failures and our management problems.

Secondly, the general feeling that there must be a grand overall strategy and uniformity of approach, underlying many of the discussions of the mid-1970s, has given way to looking at which parts of

Systematics Association Special Volume No. 26, "Databases in Systematics", edited by R. Allkin and F. A. Bisby, 1984, pp. 321–324. Academic Press, London and Orlando.

DATABASES IN SYSTEMATICS
ISBN 0 12 053040 6

systematics can be aided in the short and medium term, with the huge range of micros, minis and mainframes now available. Ten years ago, when there were really only big and somewhat clumsy machines available, the thinking was, quite naturally, that big problems should be addressed. This in effect meant important smaller problems were left unsolved. In itself this brought the limitation that only a few of the richer institutions felt they could afford to address these so-called major tasks. Sadly, the final outcome was that almost no results of any importance have been forthcoming.

The coming of the mini and the micro with ready-made software has allowed hundreds of institutions with little support staff to attack those smaller but equally important tasks. It has given us a new way of looking at, and provided many new ideas as to how we may cope with the major tasks, as well as how to handle them in smaller and more manageable portions.

The third area of change is that the machines and their software are far more user friendly and allow a far wider range of biologists, with only a relatively small computer background, to resolve their particular problems. The ability to build up data from different machines into larger databases is developing well. The option remains open to resolve our own particular problems and yet we can look forward to where, if we wish, particular sets can be interrelated to build into larger more comprehensive databases. These discrete problem-solving systematic projects, which have a defined target in a relatively short time-scale of years rather than decades, means that output can be produced at the end of the project. This in itself will help to satisfy funding agencies. It will also help to reassure ourselves and our masters of the ease with which we should be able to produce appropriate papers for publication but, more importantly from my point of view, they can often also be combined to resolve larger problems.

We have replaced the massive coding exercises needed for the earlier machines by the use of "real" words put in with a standard typewriter keyboard where the machine deals with all the encoding. We now have a tool for all scientific personnel to be able to key in data in formatted programmes which will help resolve quite simple problems. For example, collecting labels, which can prepare for the production of itinerary and eventually species listings at the same time when the botanist or his assistant keys in the identifications. The very simplicity of modern equipment allows us a great opportunity to expand our curatorial capacity with little increase in staff. A very important feature today!

I do not intend to deal with the diversity of problem-solving capacities now also available: that is self-evident from the other conference papers. It may however be useful to identify two areas in which I see very important developments affecting taxonomists such as those in my own institution at Kew.

The first concerns the vast amount of taxonomic data we produce which goes through so many routine processes before it is published, many of which are totally outside the control of the author or the editor. With word processors today the paper, format, proof-reading and text handling can all be carried out "in house" before going to the printer. This not only saves by not including errors created by others and therefore having to look for them later, but it reduces them to a one-off correcting exercise, so dramatically reducing the checking and proof-reading time; it also heavily reduces production costs and thus final selling price. What is the point of carrying out research if it cannot be made available quickly and cheaply? Many different formats and combinations of print-outs can be made up from the stored text allowing one to fulfil different types of contracts, all drawing on the same database.

Another important development at Kew is the conversion of earlier traditional databases into more accessible and more readily compatible forms. Here I should say something about the *Index Kewensis*. The approximately two million species references are available separately in the original two volumes and many supplements; or, if you are very lucky, cut and pasted in supplement order under the generic name as at Kew. Original calculations some nine years ago showed that it would take 80 labour years and some £300 000 to key this data in, proof-read it, etc. Today, I am glad to say, with equipment run by the Department of Health and Social Security in Newcastle and with a great deal of support in the Ministry of Agriculture and via our own two ladies in the *Index Kewensis* team, we hope by the end of 1983 to have all the data captured for reading on Kew's own computer, also promised for the autumn of 1983. We envisage a service to follow, when we have cleared the problems, which will allow for family or generic listings to be made on demand, possibly on-line or as a tape service. The production of printed volumes is appropriate where there is demand. More vital to some will be the fact that maybe yearly *Index Kewensises* can be made available for future listings, or that the latest data keyed in will be available in a print-out of species in any particular genus.

The future, I believe, looks very rosy, despite financial and manpower limitations. I believe we now have the tools available to resolve many of the problems which ten years ago were impossible dreams.

REFERENCE

Brenan, J. P. M., Ross, R. and Williams, J. T. (eds) (1975). "Computers in Botanical Collections". Plenum Press, London and New York.

Index of Key Words

A

Addresses, postal, database of, 100
Almost flat files, 53, 57, 61, *see* Bumpy Files
Amanita bisporigera, 251
Amanita muscaria, 251
Amanita virosa, 251
Amphibian taxonomic directory, 107
Anthocyanidins, 209
Arabidopsis, 224
ARCBASE, 120
Asteraceae, 312
AUTOGLOBAL checking program, 146
Automatic
  identification, 245–246, 249–261
  mapping, 245
  typesetting, 280, 284–287
Avian Taxonomic Directory, 107

B

Basidiospores, 250, 251, 255
Bibliographic
  database, 100, 101, 125, 128, 221
  file, 73
BIOSIS, 69–70, 221
Biosystematics, 79, 81, 82, 87
BIOPT4, 224
Biotopomaps, 224
Boolean expressions, 55–61, 182, 213
Botanic gardens, 240
Botanical terminology, 169, 265–267, 309–320
Bracteoles, 317, 318
Bracts,
  definition of, 317
  description of, 317, 318
  types of, 317
BRASS BAND, 219, 220–228, 231
Brassicaceae, 219–233
Bumpy files, 63, 273, *see* Flat file data structures

C

Cactaceae, 307
Camera-ready copy, 284, *see* Typesetting
Canavanine, 203–207
Cap
  scalp, 250
  trama, 250, 251
Cardamine, 225–226
Catalogues, 131–132, 240– 241
Character, *see also* Descriptor
  definition, 171–172, 266, 267, 275
  dependance, 263, 271–273
  hierarchy, 302
  master, 274
  selection, 275–276
  sister, 263, 274
  states, 182, 264, 266–270, 274
Chemical database file, 179
Chemotaxonomy, 201, 209, 210, 217
Cladistics, 13, 27
Classification, 17, 18–19, 21, 190–191, 196–199, 210, 271, 275–276
Cluster analysis, 216
CODASYL, 38, 42
Coding, 279
COLLECTOR data sets, 142, 146, 150
Complexed variation, 263, 266, 269

Computer languages,
 extensibility, 295–297
 Forth, 291, 293, 296–300, 302, 303–306
 interactive command, 295
CONFOR, 27, 175, 179, 182, 187, 264, 282–283, 284–286, 288
Conservation, 91, 92
Conversion, 279, 282–283, 302
Corsica, 239
Cruciferae, 219–233
Cultivar, 189, 190–199
CURARBA, 120
Customers, 18, 19, 21–25, 29
Cyme, helicoid, 315
Cystidia, 251, 253

D

Data
 coding of, 160
 file, 73, 179
 Manipulation Language (DML), 39
 processing, 1
 protection, 304
Database, 35
 architecture, 36, 263, 291
 bibliographic, 100, 101, 125, 126, 128, 221
 chemical, 201–208
 chemotaxonomic, 179, 209, 210, 217
 curatorial, 156–163, 165, 166, 167, 172, 225
 design of, 26, 27, 162, 270, 271, 273–274
 descriptive, 26, 189, 264, 265, 267, 294
 distributed, 113, 122
 floristic, 239, 309, 319
 general purpose systems, 293
 herbarium, see curatorial
 implementation languages, 295, 298
 management of, 161–162
 management systems (DBMS), 46–52, 263, 291, 292–297, 300
 revision, 17, 28–29, 185
 small special purpose, 294, 297, 298
DELTA format, 28, 182–183 268, 269, 272, 280–282, 285
DESC, 171, 275
Description, 168–169, 170, 187, 263, 279, 280, *see also* Character
Descriptive taxonomic data, 18, 21, 22, 26, 183, 264, 265 267, 294
Descriptor, 54–57, 59–62, *see also* Character
 groups, 62–63
 independant, 62, *see also* Character, dependance
 multiple, 63–65
 states, 55, 59
 types, 267
Diffusion model, 20
Distinctness, 189–191, 198, 199
Distributed database, 113, 122
Distribution maps, 113, 115, 120, 125, 133, 135, 224, 245
DMS-INQUIRY, 137, 143, 146
Documentation, 79, 82
DOKUM, 128, 129
Double reporting, 106

E

Error checking, 272, 293
Euphorbia peplus, 313, 314, 318, 319
Europe, 79, 80, 223
 floristic cartography, 113, 115
EXIR, 27, 28, 175, 179, 182, 184, 185, 194–196, 198, 209, 210, 213, 264, 268, 272, *and see* TAXIR
EXIRPOST, 27, 175, 184–185

F

Fact documentation, 125, 126, 127, 130, 135
Fagus sylvatica, 317
Fascicles, 31, 168–169, *see also* Flora and Fauna
FASTDIZ, 120
Fauna, 17, 18, 19, 263, 271
Field guides, 22, 24, 30, 265, 266
Fixed format, 158
Flat file data structures, 55, 273, *see* Almost Flat Files

Flavonols, 210, 213
Flora, 17, 18, 19, 24, 29, 157, 159, 240, 263, 271
  Europaea, 22, 24, 31, 79, 80, 81, 83, 85, 223
  of Italy, 116, 117
  local, Index of, 240
  of North America, 2, 225, 223, 266–267
  de Veracruz, 25, 156, 157, 160, 161, 165–174, 264
Floristic cartography, 113, 115
Floristics, 6, 79–82, 222
Format, 280–282
  conversion, 302
Forth language system 291, 293, 296–306
Friuli-Venezia Giulia, 121

G

Genes, 197, 198
Genetics, 189, 194, 197, 198, 199
Geneva, 236
GENSPEC data sets, 141, 142, 146, 149
Geobotany, 113, 117
GEOECOLOGY, 228
Geographical, *see* Distribution maps
  coding schemes, 94
  database file, 179, 185
  distributions, 133
  institutional codes, 107–108
Geranium molle, 313, 318
Glossary of botanical terms, 169
Graphs, directed, 271
Grasses, 279, 288
Greene Herbarium, 220, 225

H

Hardware, 44, 46, 47, 51
Herbarium, 6, 8, 9, 10, *see also* Database, curatorial
  and library management, 235, 236, 238–239
  computerization of, 113, 121
  organization of, 155
  procedures, 151
  specimen, annotation, 161
  specimen labels, 139, 142, 149
Hierarchic design, 35, 37–38
Hierarchical relationship among taxa, 269, 271, 273
Hierarchy
  file, 73
  taxonomic, 269, 271, 273, 291, 301, 302, 303,
Hymenophoral trama, 250, 251
Hylocereinae, 307

I

Identification, 18, 19, 21, 22, 27, 106, 170, 171, 245–246, 249–261, 272
IDMS, 38
IMS, 37
Independant descriptors, 62, *see* Character, dependance
Index Kewensis, 7, 321, 323
Indiana, 222
Indigofera, 203–208
Indospicine, 203–204, 205
Inflorescence,
  classification of, 311, 315–316, 319
  definition of, 311–312
  description of, 309, 310–311, 312–316, 319
  development of, 315
  evolution of, 315
INFOL, 240–242, 245
Information,
  centre, 175, 176–177, 187
  processing, 2, 3, 5
  retrieval, 27, 74, 182, 213, 263, 264, 267, 268, 271, 295, 302, 307
  services, 17, 18, 73, 275
Interactive
  data entry, 291
  identification, 171, 245–246, 249–261
  retrieval, 74, 213, 263, 267, 272, 295, 302, 307
Interface levels, 293, 303
International Species Inventory System (ISIS), 103–112
ISIS taxonomic codes, 95–96, 98

Italy, 113
Items, 55

J

Join, 41, 66

K

Keys, 186, 272, 274, 279, 283, *and see* Identification

L

Lactarius tabidus, 251
Lathyrine, 211
Lathyrus, 177, 180, 184, 186, 211, 214, 216
Leguminosae, 31, 176, 177, 186, 201, 209, 210, *see also Pisum sativum*
Lens, 177, 211, 216
LINKAGE, 179, 216
LIT BRASS, 221
Literature database, 126, *see* Database, bibliographic
Local Floras Index, 240
Loranthaceae, 314

M

Mainframes, 35, 322
Malva sylvestris, 311, 317
Mammalian Taxonomic Directory, 107
MAPPA, 121
Mapping, *see* Distribution maps
  automatic, 245,
Medcheck list, 239
Mediterranean area, 239
Mexico, 172
Microcomputer, 43, 44, 45, 48, 203, 297, 321, 322
Microfiche, 284
Minicomputer, 322
Monograph, 17, 18, 19, 21, 24, 29, 175, 263, 264, 271
Morphological database file, 178, 179, 181–183
Multipurpose taxonomic database, 25–27, 273
Multiple descriptors, 63–65
Multistage decision process, 230–231
Mushrooms,
  toxic, 249, 256–257

N

National list, 189, 190
National parks, 99
Network, 46, 48
  design, 35, 38–40, 42
Nomenclature, 17, 18–19, 21
  file, 72, 179, 180, 185
Non-protein amino acid, 201, 202, 203, 207, 209–211, 213
North America, 223, *see also* Flora of
Numerical taxonomy, 11

O

Operating system, 292, 293
On-line, *see* Interactive

P

Panicle, 310, 312, 318
Papilionoideae, 175–187, 203, 205
Paraguay, 239
Pattern recognition, 208
Peru, 239
Phytoalexins, 209, 210, 216
Phytogeography, 113, 117
Phytosociology, 113, 114, 116, 118
Pisum, 117, 211, 216
Pisum sativum, 189, 194
Plant breeders rights, 189, 190
Plant toxins, 203–204
Pointer structure, 64–66
Polyclaves, 171
Portable software, 299
PRECIS, 156, 157, 159–162
PRECIS I, 137, 138–140, 150
PRECIS II, 140–153
Prejoin, 65, 66
PRENO, 141, 143, 149, 151
Probabilistic information, 270
Programa Flora, 157, 159
Programming, 44, 47, 50
Protected areas, 92, 98–99
Protein amino acid, 201, 202, 207

Q

Quality control, 146
Questionnaire, 159
  program, 310

R

Raceme, 312, 314, 315, 318
REBUS, 238
REGLOC data sets, 142–143
Relational, 47, 48, 49, 52
  database model, 66, 273
  design, 40–41, 42
Remote access, 74
Reptile Taxonomic Directory, 107
Retail model, 18, 19
Retrieval, 27, 74, 182, 213, 263, 264, 267, 268, 271, 295, 302, 307
Rubus, 310

S

Secondary
  compounds, 202
  substances, 209
Secondary information services, 73
SELGEM, 161, 162
SIBIL, 239
Single link cluster analysis, 216
Sister characters, 263, 274
Software, 43, 46–48, 50, 52
  portable, 299
Soviet Union, 223–224
SPECIES PER GRID program, 143, 152
Specimen
  data sets, 140–141
  encoding form, 141
  herbarium, annotation, 161
  identification, 171, 245–246, 249–261
SPECTAGS, 149, 150
Standardization, 113, 114, 115, 118 121–122
Standard taxonomies, 106
SYMVU, 223
SYNONYMS, 176, 179, 180
Systematics, 2, 8, 9, 13, 14, 43, 219

T

Tailor-made products, 175, 186
TANAIDACEA database, 126, 127, 133
Taxa-tree, 96–97
TAXIR, 26, 53–61, 117, 157, 161, 162, *and see* EXIR
Taxonomic
  codes, 107–109
  coding schemes, 95–96
  descriptive data, 18, 21, 22, 26, 264, 265, 267, 294
  descriptive data, quality of, 265, 310
  directories, 107
  hierarchy, 271, 273, 291, 301, 302, 303
  information services, 17, 18, 21, 24, 25, 28
  products, 18–21, 23, 31
  profession, role of, 18–21, 30
  references file, 69, 71, 72, 73–76
Taxonomy, 2, 4, 5, 9, 10, 13, 14, 79, 80, 82
TELDOK, 129
TESAURO, 157, 160, 161, 162
Text processing, 235, 239
Thelypodieae, 26
Threatened
  animals, 91, 92, 95, 98
  plants, 91, 92, 95, 97
Toxin, 256, 257
Translation, 169, 171, 282–283
Tree structures, 37–38, 91, 97, 271–272
Trees in urban area, 240
Trieste code, 114
TUBIBUE 126–127
Typesetting, 280, 284–287

V

Variable
  observations, 302
  taxa, 268, 269, 273
Variety, *see* Cultivar
Veracruz, 25, 156, 157, 159, 160, 161, 165–174, 264
Vicia, 22, 177, 186, 215, 216
Vicieae, 26, 30, 175–187, 210
Viscum album, 312

W

Weeds, 240
Word processors, 323

Z

Zoological record, 69–70

Editor-in-Chief, Special Volume Series

D. L. HAWKSWORTH PhD DSc FLS FIBiol

Commonwealth Mycological Institute, Kew

Systematics Association Publications

1. BIBLIOGRAPHY OF KEY WORKS FOR THE IDENTIFICATION OF THE BRITISH FAUNA AND FLORA *3rd edition* (1967)
   Edited by G. J. Kerrich, R. D. Meikle and N. Tebble
   Out of print
2. FUNCTION AND TAXONOMIC IMPORTANCE (1959)
   Edited by A. J. Cain
3. THE SPECIES CONCEPT IN PALAEONTOLOGY (1956)
   Edited by P. C. Sylvester-Bradley
4. TAXONOMY AND GEOGRAPHY (1962)
   Edited by D. Nichols
5. SPECIATION IN THE SEA (1963)
   Edited by J. P. Harding and N. Tebble
6. PHENETIC AND PHYLOGENETIC CLASSIFICATION (1964)
   Edited by V. H. Heywood and J. McNeill
   Out of print
7. ASPECTS OF TETHYAN BIOGEOGRAPHY (1967)
   Edited by C. G. Adams and D. V. Ager
8. THE SOIL ECOSYSTEM (1969)
   Edited by H. Sheals
9. ORGANISMS AND CONTINENTS THROUGH TIME (1973)†
   Edited by N. F. Hughes

Published by the Association

Systematics Association Special Volumes

1. THE NEW SYSTEMATICS (1940)
   Edited by Julian Huxley (Reprinted 1971)
2. CHEMOTAXONOMY AND SEROTAXONOMY (1968)*
   Edited by J. G. Hawkes
3. DATA PROCESSING IN BIOLOGY AND GEOLOGY (1971)*
   Edited by J. L. Cutbill
4. SCANNING ELECTRON MICROSCOPY (1971)*
   Edited by V. H. Heywood
5. TAXONOMY AND ECOLOGY (1973)*
   Edited by V. H. Heywood
6. THE CHANGING FLORA AND FAUNA OF BRITAIN (1974)*
   Edited by D. L. Hawksworth
7. BIOLOGICAL IDENTIFICATION WITH COMPUTERS (1975)*
   Edited by R. J. Pankhurst
8. LICHENOLOGY: PROGRESS AND PROBLEMS (1976)*
   Edited by D. H. Brown, D. L. Hawksworth and R. H. Bailey
9. KEY WORKS TO THE FAUNA AND FLORA OF THE BRITISH ISLES AND NORTHWESTERN EUROPE (1978)*
   Edited by G. J. Kerrich, D. L. Hawksworth and R. W. Sims

*Published by Academic Press for the Systematics Association
†Published by the Palaeontological Association in conjunction with the Systematics Association

10. MODERN APPROACHES TO THE TAXONOMY OF RED AND BROWN ALGAE (1978)★
Edited by D. E. G. Irvine and J. H. Price
11. BIOLOGY AND SYSTEMATICS OF COLONIAL ORGANISMS (1979)★
Edited by G. Larwood and B. R. Rosen
12. THE ORIGIN OF MAJOR INVERTEBRATE GROUPS (1979)★
Edited by M. R. House
13. ADVANCES IN BRYOZOOLOGY (1979)★
Edited by G. P. Larwood and M. B. Abbot
14. BRYOPHYTE SYSTEMATICS (1979)★
Edited by G. C. S. Clarke and J. G. Duckett
15. THE TERRESTRIAL ENVIRONMENT AND THE ORIGIN OF LAND VERTEBRATES (1980)★
Edited by A. L. Panchen
16. CHEMOSYSTEMATICS: PRINCIPLES AND PRACTICE (1980)★
Edited by F. A. Bisby, J. G. Vaughan and C. A. Wright
17. THE SHORE ENVIRONMENT: METHODS AND ECOSYSTEMS (2 Volumes) (1980)★
Edited by J. H. Price, D. E. G. Irvine and W. F. Farnham
18. THE AMMONOIDEA (1981)★
Edited by M. R. House and J. R. Senior
19. BIOSYSTEMATICS OF SOCIAL INSECTS (1981)★
Edited by P. E. Howse and J.-L. Clément
20. GENOME EVOLUTION (1982)★
Edited by G. A. Dover and R. B. Flavell
21. PROBLEMS OF PHYLOGENETIC RECONSTRUCTION (1982)★
Edited by K. A. Joysey and A. E. Friday
22. CONCEPTS IN NEMATODE SYSTEMATICS (1983)★
Edited by A. R. Stone, H. M. Platt and L. F. Khalil
23. EVOLUTION, TIME AND SPACE: THE EMERGENCE OF THE BIOSPHERE (1983)★
Edited by R. W. Sims, J. H. Price and P. E. S. Whalley
24. PROTEIN POLYMORPHISM: ADAPTIVE AND TAXONOMIC SIGNIFICANCE (1983)★
Edited by G. S. Oxford and D. Rollinson
25. CURRENT CONCEPTS IN PLANT TAXONOMY (1984)★
Edited by V. H. Heywood and D. M. Moore
26. DATABASES IN SYSTEMATICS (1984)★
Edited by R. Allkin and F. A. Bisby

In preparation

27. SYSTEMATICS OF THE GREEN ALGAE (1984)★
Edited by D. E. J. Irvine and D. M. John

★Published by Academic Press for the Systematics Association